Real Analysis:
A Beginner's Guide

Real Analysis: A Beginner's Guide

Rakesh Goyal

Oxford Book Company

Jaipur ▪ New Delhi

Published By

Oxford Book Company

Regd. Off. "Bony Residency", Gate No. 2
Opp. Tilak Public School, Vishweswariya Nagar,
Gopalpura Road, Jaipur - 302018, India
Ph: 0141 - 2761280, Telefax: 2761381
email - oxfordbookjpr@gmail.com
website - oxfordbookcompany.com

Delhi Office: 102, 1st Floor, Satyam House,
4327/3, Ansari Road, Darya Ganj,
New Delhi - 110002, India
Ph: 011-45652440

Distributed By

KEYA INTERNATIONAL PRESS LLP
E-mail: keyainternationalpressllp@gmail.com

First published in India, 2026

ISBN : 978-93-5524-787-2

Typeset by Shivangi Computers, Jaipur

Printed and bound in India by Thomson Press Ltd.,
New Delhi

Contents

Preface

In mathematics, analysis is the field that deals with limits and inequalities. In this course, the most fundamental ideas in analysis are covered. Establishing a solid foundation for calculus of one variable and introducing the reader to rigorous proofs in analysis are the two main objectives of the course. The purpose of calculus is to prepare you, the student, to use mathematics without explaining the validity of what you have studied. To effectively utilize or teach mathematics, you need to understand why something is true rather than just knowing it is true. Discover why calculus is true in this course. The purpose of this is to help you grasp the concepts of a limit, derivative, and integral. I'll give you an example. Simple operations like charging the battery, fixing broken headlights, and changing the oil are all that an auto mechanic can perform. It will be impossible for him to diagnose and resolve issues on his own. Not knowing the definition of the Riemann integral, a high school teacher will be unable to adequately respond to all of the questions that students may have.

The study of real numbers and real valued functions of a real variable is known as real analysis (or, more conventionally, the theory of functions of a real variable). In specifically, it covers the calculus of real numbers, continuity, smoothness, and related features of real-valued functions, as well as the analytic aspects of real functions and sequences, such as convergence and limits of real number sequences. The field of analysis known as "real analysis" examines ideas like sequences and their boundaries, continuity, differentiation, integration, and function sequences. Since the extended real line is made up of both positive and negative infinity, real analysis is by definition concerned with the real numbers.

A number of significant lattice-theoretic characteristics that are lacking in complex numbers are present in real numbers. Above all, the real numbers constitute an ordered field in which multiplication and addition maintain positivity. However, the real numbers have the least upper bound attribute and are ordered in a total manner. The mean value theorem, the intermediate value theorem, and the monotone convergence theorem are just a few of the significant real-analysis outcomes that arise from these order-theoretic qualities.

Real analysis conclusions are expressed for real numbers, but many of these findings are applicable to other mathematical objects. Riesz space and positive operator theories are two examples of the many concepts in functional analysis and operator theory that generalize features of the real numbers. Mathematicians also evaluate operator sequences pointwise or give consideration to real and imaginary components of complicated sequences.

This book, suitable for elementary calculus students, covers functions, differentiation, integration, series of real numbers, uniform convergence, and algebraic structure of number systems. It covers fundamentals and rigor, making it ideal for classroom or independent study.

— ***Rakesh Goyal***

Chapter 1

Uniform Continuity

In mathematics, a function f is uniformly continuous if, roughly speaking, it is possible to guarantee that $f(x)$ and $f(y)$ be as close to each other as we please by requiring only that x and y are sufficiently close to each other; unlike ordinary continuity, the maximum distance between x and y cannot depend on x and y themselves. For instance, any isometry (distance-preserving map) between metric spaces is uniformly continuous.

Every uniformly continuous function between metric spaces is continuous. Uniform continuity, unlike continuity, relies on the ability to compare the sizes of neighbourhoods of distinct points of a given space. In an arbitrary topological space this may not be possible. Instead, uniform continuity can be defined on a metric space where such comparisons are possible, or more generally on a uniform space.

The equicontinuity of a set of functions is a generalization of the concept of uniform continuity.

Every continuous function on a compact set is uniformly continuous.

Definition for Functions on Metric Spaces

Given metric spaces (X, d_1) and (Y, d_2), a function $f : X \to Y$ is called uniformly continuous if for every real number $\varepsilon > 0$ there exists $\delta > 0$ such that for every $x, y \in X$ with $d_1(x, y) < \delta$, we have that $d_2(f(x), f(y)) < \varepsilon$.

If X and Y are subsets of the real numbers, d_1 and d_2 can be the standard Euclidean norm, $|| \cdot ||$, yielding the definition:

for all $\varepsilon > 0$ there exists a $\delta > 0$ such that for all $x, y \in X$, $|x - y| < \delta$ implies $|f(x) - f(y)| < \varepsilon$.

The difference between being uniformly continuous, and simply being continuous at every point, is that in uniform continuity the value of δ depends only on ε and not on the point in the domain.

Local Continuity Versus Global Uniform Continuity

Continuity itself is a *local* (more precisely, *pointwise*) property of a function—that is, a function f is continuous, or not, at a particular point. When we speak of a function being continuous on an interval, we mean only that it is continuous at each point of the interval. In contrast, uniform continuity is a *global* property of f, in the sense that the standard definition refers to *pairs* of points rather than individual points. On the other hand, it is possible to give a definition that is *local* in terms of the natural extension f^* (the characteristics of which at nonstandard points are determined by the global properties of f), although it is not possible to give a local definition of uniform continuity for an arbitrary hyperreal-valued function.

The mathematical statements that a function is continuous on an interval I and the definition that a function is uniformly continuous on the same interval are structurally very similar. Continuity of a function for every point x of an interval can thus be expressed by a formula starting with the quantification

$$\forall x \forall \varepsilon \exists \delta \forall y (| y - x | < \delta \Rightarrow | f(y) - f(x) | < \varepsilon),$$

which is equivalent to

$$\forall \varepsilon \forall x \exists \delta \forall y (| y - x | < \delta \Rightarrow | f(y) - f(x) | < \varepsilon),$$

whereas for uniform continuity, the order of the second and third quantifiers is reversed:

$$\forall \varepsilon \exists \delta \forall x \forall y (| y - x | < \delta \Rightarrow | f(y) - f(x) | < \varepsilon)$$

(the domains of the variables have been deliberately left out so as to emphasize quantifier order). Thus for continuity at each point, one takes an arbitrary point x, and then there must exist a distance δ,

$$\cdots \forall x \exists \delta \cdots,$$

while for uniform continuity a single δ must work uniformly for all points x (and y):

$$\cdots\exists\delta\forall x\forall y\cdots.$$

Examples

- Every Lipschitz continuous map between two metric spaces is uniformly continuous. In particular, every function which is differentiable and has bounded derivative is uniformly continuous. More generally, every Hlder continuous function is uniformly continuous.
- Every member of a uniformly equicontinuous set of functions is uniformly continuous.
- The tangent function is continuous on the interval ($-\pi/2$, $\pi/2$) but is *not* uniformly continuous on that interval.
- The exponential function $x \mapsto e^x$ is continuous everywhere on the real line but is not uniformly continuous on the line.

Properties

Every uniformly continuous function is continuous, but the converse does not hold. Consider for instance the function $f:\mathbb{R}\to\mathbb{R}, x\mapsto x^2$. Given an arbitrarily small positive real number $\in$, uniform continuity requires the existence of a positive number δ such that for all x_1, x_2 with $|x_1 - x_2| < \delta$, we have $|f(x_1) - f(x_2)| < \epsilon$. But

$$f(x+\delta) - f(x) = 2x\delta + \delta^2 = \delta(2x+\delta),$$

and for all sufficiently large x this quantity is greater than $\in$.

The image of a totally bounded subset under a uniformly continuous function is totally bounded. However, the image of a bounded subset of an arbitrary metric space under a uniformly continuous function need not be bounded: as a counterexample, consider the identity function from the integers endowed with the discrete metric to the integers endowed with the usual Euclidean metric.

The Heine–Cantor theorem asserts that every continuous function on a compact set is uniformly continuous. In particular, if a function is continuous on a closed bounded interval of the real line, it is uniformly continuous on that interval. The Darboux integrability of continuous functions follows almost immediately from the uniform continuity theorem.

If a real-valued function f is continuous on $[0,\infty)$ and $\lim_{x\to\infty} f(x)$ exists (and is finite), then f is uniformly continuous. In particular, every element of $C_0(\mathbb{R})$, the space of continuous functions on ¡that vanish at infinity, is uniformly continuous. This is a generalization of the Heine-Cantor theorem mentioned above, since $C_c(\mathbb{R}) \subset C_0(\mathbb{R})$.

History

The first published definition of uniform continuity was by Heine in 1870, and in 1872 he published a proof that a continuous function on an open interval need not be uniformly continuous. The proofs are almost verbatim given by Dirichlet in his lectures on definite integrals in 1854. The definition of uniform continuity appears earlier in the work of Bolzano where he also proved that continuous functions on an open interval do not need to be uniformly continuous. In addition he also states that a continuous function on a closed interval is uniformly continuous, but he does not give a complete proof.

Other Characterizations

Non-standard analysis: In non-standard analysis, a real-valued function f of a real variable is microcontinuous at a point a precisely if the difference $f^*(a + \delta)$ " $f^*(a)$ is infinitesimal whenever δ is infinitesimal. Thus f is continuous on a set A in R precisely if f^* is microcontinuous at every real point a " A. Uniform continuity can be expressed as the condition that (the natural extension of) f is microcontinuous not only at real points in A, but at all points in its non-standard counterpart (natural extension) $^*A$ in *R. Note that there exist hyperreal-valued functions which meet this criterion but are not uniformly continuous, as well as uniformly continuous hyperreal-valued functions which do not meet this criterion, however, such functions cannot be expressed in the form f^* for any real-valued function f.

Characterization Via Sequences

For a function between Euclidean spaces, uniform continuity can be defined in terms of how the function behaves on sequences (Fitzpatrick 2006). More specifically, let A be a subset of R^n. A function $f: A \to R^m$ is uniformly continuous if and only if for every pair of sequences x_n and y_n such that

$$\lim_{n\to\infty} |x_n - y_n| = 0$$

we have

$$\lim_{n\to\infty} |f(x_n) - f(y_n)| = 0.$$

Relations with the Extension Problem

Let X be a metric space, S a subset of X, and $f: S \to R$ a continuous function. When can f be extended to a continuous function on all of X?

If S is closed in X, the answer is given by the Tietze extension theorem: always. So it is necessary and sufficient to extend f to the closure of S in X: that is, we may assume without loss of generality that S is dense in X, and this has the further pleasant consequence that if the extension exists, it is unique.

Let us suppose moreover that X is complete, so that X is the completion of S. Then a continuous function $f: S \to R$ extends to all of X if and only if f is Cauchy-continuous, i. e., the image under f of a Cauchy sequence remains Cauchy. (In general, Cauchy continuity is necessary and sufficient for extension of f to the completion of X, so is *a priori* stronger than extendability to X.)

It is easy to see that every uniformly continuous function is Cauchy-continuous and thus extends to X. The converse does not hold, since the function $f: R \to R, x \mapsto x^2$ is, as seen above, not uniformly continuous, but it is continuous and thus — since R is complete — Cauchy continuous. In general, for functions defined on unbounded spaces like R, uniform continuity is a rather strong condition. It is desirable to have a weaker condition from which to deduce extendability.

For example, suppose $a > 1$ is a real number. At the precalculus level, the function $f : x \mapsto a^x$ can be given a precise definition only for rational values of x (assuming the existence of qth roots of positive real numbers, an application of the Intermediate Value Theorem). One would like to extend f to a function defined on all of R. The identity

$$f(x+\delta) - f(x) = a^x(a^\delta - 1)$$

shows that f is not uniformly continuous on all of Q; however for any bounded interval I the restriction of f to $Q \cap I$ is uniformly continuous, hence Cauchy-continuous, hence f extends to a continuous function on I. But since this holds for every I, there is then a unique extension of f to a continuous function on all of R.

More generally, a continuous function $f : S \to R$ whose restriction to every bounded subset of S is uniformly continuous is extendable to X, and the converse holds if X is locally compact.

A typical application of the extendability of a uniform continuous function is the proof of the inverse Fourier transformation formula. We first prove that the formula is true for test functions, there are densely many of them. We then extend the inverse map to the whole space using the fact that linear map is continuous; thus, uniformly continuous.

Generalization to Topological Vector Spaces

In the special case of two topological vector spaces $V$and W, the notion of uniform continuity of a map $f : V \to W$ becomes: for any neighbourhood $B$of zero in W, there exists a neighbourhood A of zero V in such that $v_1 - v_2 \in A$ implies $f(v_1) - f(v_2) \in B$.

For linear transformations $f : V \to W$, uniform continuity is equivalent to continuity. This fact is frequently used implicitly in functional analysis to extend a linear map off a dense subspace of a Banach space.

Generalization to Uniform Spaces

Just as the most natural and general setting for continuity is topological spaces, the most natural and general setting for the study of *uniform* continuity are the uniform spaces. A function $f : X \to Y$ between uniform space is called *uniformly continuous* if for every entourage V in Y there exists an entourage U in X such that for every (x_1, x_2) in U we have $(f(x_1), f(x_2))$ in V.

In this setting, it is also true that uniformly continuous maps transform Cauchy sequences into Cauchy sequences and that continuous maps on compact uniform spaces are automatically uniformly continuous.

Each compact Hausdorff space possesses exactly one uniform structure compatible with the topology. A consequence is a generalisation of the Heine-Cantor theorem: each continuous function from a compact Hausdorff space to a uniform space is uniformly continuous.

Modulus of Continuity

In mathematical analysis, a modulus of continuity is a function ω : [0, ¥] → [0,¥] used to measure quantitatively the uniform continuity of functions. So, a function $f : I \to \mathrm{R}$ admits ω as a modulus of continuity if and only if

$$|f(x) - f(y)| \leq \omega(|x - y|),$$

for all x and y in the domain of f. Since moduli of continuity are required to be infinitesimal at 0, a function turns out to be uniformly continuous if and only if it admits a modulus of continuity. Moreover, relevance to the notion is given by the fact that sets of functions sharing the same modulus of continuity are exactly equicontinuous families. For instance, the modulus $\omega(t) := kt$ describes the k-Lipschitz functions, the moduli $\omega(t) := kt^{\alpha}$ describe the Hölder continuity, the modulus $\omega(t) := kt(|\log(t)|+1)$ describes the almost Lipschitz class, and so on. In general, the role of ω is to fix some explicit functional dependence of ε on δ in the (ε, δ) definition of uniform continuity. The same notions generalize naturally to functions between metric spaces. Moreover, a suitable local version of these notions allows to describe quantitatively the continuity at a point in terms of moduli of continuity.

A special role is played by concave moduli of continuity, especially in connection with extension properties, and with approximation of uniformly continuous functions. For a function between metric spaces, it is equivalent to admit a modulus of continuity that is either concave, or subadditive, or uniformly continuous, or sublinear (in the sense of growth). Actually, the existence of such special moduli of continuity for a uniformly continuous function is always ensured whenever the domain is either a compact, or a convex subset of a normed space. However, a uniformly continuous function on a general metric space admits a concave modulus of continuity if and only if the ratios

$$\frac{d_Y(f(x),f(x'))}{d_X(x,x')}$$

are uniformly bounded for all pairs $(x, x2)$ bounded away from the diagonal of X. The functions with the latter property constitute a special subclass of the uniformly continuous functions, that in the following we refer to as the *special uniformly continuous* functions. Real-valued special uniformly continuous functions on the metric space X can also be characterized as the set of all functions that are restrictions to X of uniformly continuous functions over any normed space isometrically containing X. Also, it can be characterized as the uniform closure of the Lipschitz functions on X.

Formal Definition

Formally, a modulus of continuity is any real-extended valued function $\omega : [0, ¥] \to [0, ¥]$, vanishing at 0 and continuous at 0, that is

$$\lim_{t\to 0} \omega(t) = \omega(0) = 0.$$

Moduli of continuity are mainly used to give a quantitative account both of the continuity at a point, and of the uniform continuity, for functions between metric spaces, according to the following definitions.

A function $f : (X, d_X) \to (Y, d_Y)$ admits ω as (local) modulus of continuity at the point x in X if and only if,

$$\forall x' \in X : d_Y(f(x),f(x')) \le \omega(d_X(x,x')).$$

Also, f admits ω as (global) modulus of continuity if and only if,

$$\forall x,x' \in X : d_Y(f(x),f(x')) \le \omega(d_X(x,x')).$$

One equivalently says that ω is a modulus of continuity (resp., at x) for f, or shortly, f is ω-continuous (resp., at x). Here, we mainly treat the global notion.

Elementary Facts

- If f has ω as modulus of continuity and $\omega_1 \geq \omega$, then, obviously, f admits ω_1 too as modulus of continuity.
- If $f : X \to Y$ and $g : Y \to Z$ are functions between metric spaces with moduli respectively ω_1 and ω_2, then the composition map $g \circ f : X \to Z$ has modulus of continuity $\omega_2 \circ \omega_1$.
- If f and g are functions from the metric space X to the Banach space Y, with moduli respectively ω_1 and ω_2, then any linear combination $af+bg$ has modulus of continuity $|a|\omega_1+|b|\omega_2$. In particular, the set of all functions from X to Y that have ω as a modulus of continuity is a convex subset of the vector space $C(X, Y)$, closed under pointwise convergence.
- If f and g are bounded real-valued functions on the metric space X, with moduli respectively ω_1 and ω_2, then the pointwise product fg has modulus of continuity $\|g\|_\infty \omega_1 + \|f\|_\infty \omega_2$.
- If $\{f_\lambda\}_{\lambda \in \Lambda}$ is a family of real-valued functions on the metric space X with common modulus of continuity ω, then the inferior envelope $\inf_{\lambda \in \Lambda} f_\lambda$, respectively, the superior envelope $\sup_{\lambda \in \Lambda} f_\lambda$, is a real-valued function with modulus of continuity ω, provided it is finite valued at every point. If ω is real-valued, it is sufficient that the envelope be finite at one point of X at least.

Remarks

- Some authors require additional properties such as ω being increasing, or continuous. However, if f admits a modulus of continuity in the weaker definition above, it also admits a modulus of continuity which is increasing and infinitely differentiable in]0, ¥[. For instance, $\omega_1(t) := \sup_{s\le t}\omega(s)$ is increasing, and $\omega_1 \ge \omega$; $\omega_2(t) := \frac{1}{t}\int_t^{2t}\omega_1(s)ds$ is also continuous, and $\omega_2 \ge \omega_1$,

and a suitable variant of the preceding definition also makes ω_2 infinitely differentiable in $]0, \infty[$.

- Any uniformly continuous function admits a minimal modulus of continuity ω_f, that is sometimes referred to as *the* (optimal) modulus of continuity of *f*:

$$\omega_f(t) := \sup\{d_Y(f(x), f(x')) : x \in X, x' \in X, d_X(x, x') = t\}, \quad \forall t \ge 0.$$

Similarly, any function continuous at the point x admits a minimal modulus of continuity at x, $\omega_f(t; x)$ (*the* (optimal) modulus of continuity of f at x) :

$$\omega_f(t;x) := \sup\{d_Y(f(x), f(x')) : x' \in X, d_X(x, x') = t\}, \quad \forall t \ge 0.$$

However, these restricted notions are not as relevant, for in most cases the optimal modulus of f could not be computed explicitly, but only bounded from above (by *any* modulus of continuity of f). Moreover, the main properties of moduli of continuity concern directly the unrestricted definition.

- In general, the modulus of continuity of a uniformly continuous function on a metric space needs to take the value $+\infty$. For instance, the function $f : \mathrm{N} \to \mathrm{N}$ such that $f(n) := n^2$ is uniformly continuous with respect to the discrete metric on N, and its minimal modulus of continuity is $\omega_f(t) = +\infty$ for any positive integer t, and $\omega_f(t) = 0$ otherwise. However, the situation is different for uniformly continuous functions defined on compact or convex subsets of normed spaces.

Special Moduli of Continuity

Special moduli of continuity also reflect certain global properties of functions such as extendibility and uniform approximation. In this section we mainly deal with moduli of continuity that are concave, or subadditive, or uniformly continuous, or sublinear. These properties are essentially equivalent in that, for a modulus ω (more precisely, its restriction on $[0, \infty[$) each of the following implies the next:

- ω is concave;
- ω is subadditive;
- ω is uniformly continuous;
- ω is sublinear, that is, there are constants a and b such that $\omega(t) \le at+b$ for all t;
- ω is dominated by a concave modulus, that is, there exists a concave modulus of continuity $\tilde{\omega}$ such that $\omega(t) \le \tilde{\omega}(t)$ for all t.

Thus, for a function f between metric spaces it is equivalent to admit a modulus of continuity which is either concave, or subadditive, or uniformly continuous, or sublinear. In this case, the function f is sometimes called a *special uniformly continuous* map. This is always true in case of either compact or convex domains. Indeed, a uniformly continuous map $f : C \to Y$ defined on a convex set C of a normed space E always admits a subadditive modulus of continuity; in particular, real-valued as a function $\omega : [0, \infty[\; \to [0, \infty[$. Indeed, it is immediate to check that the optimal modulus of continuity ω_f defined above is subadditive if the domain of f is convex: we have, for all s and t:

$$\begin{aligned}\omega_f(s+t) &= \sup_{|x-x'|=t+s} d_Y(f(x), f(x')) \\ &\le \sup_{|x-x'|=t+s} \left\{ d_Y\left(f(x), f\left(x + t\frac{x-x'}{|x-x'|}\right)\right) + d_Y\left(f\left(x + t\frac{x-x'}{|x-x'|}\right), f(x')\right)\right\} \\ &\le \omega_f(t) + \omega_f(s).\end{aligned}$$

However, a uniformly continuous function on a general metric space admits a concave modulus of continuity if and only if the ratios $d_Y(f(x), f(x')) / d_X(x, x')$ are uniformly bounded for all pairs (x, $x2$) bounded away from the diagonal of X; this condition is certainly satisfied by any bounded uniformly continuous function; hence in particular, by any continuous function on a compact metric space.

Sublinear Moduli, and Bounded Perturbations from Lipschitz

A sublinear modulus of continuity can easily found for any uniformly function which is a bounded perturbations of a Lipschitz function: if f is a uniformly continuous function with modulus of continuity ω, and g is a k Lipschitz function with uniform distance r from f, then f admits the sublinear module of continuity min{$\omega(t)$, $2r+kt$. Conversely, at least for real-valued functions, any bounded, uniformly continuous perturbation of a Lipschitz function is a special uniformly continuous function; indeed more is true as shown below. Note that as an immediate consequence, any uniformly continuous function on a convex subset of a normed space has a sublinear growth: there are constants a and b such that $|f(x)| \leq a|x|+b$ for all x.

Subadditive Moduli, and Extendibility

The above property for uniformly continuous function on convex domains admits a sort of converse at least in the case of real-valued functions: that is, every special uniformly continuous real-valued function $f: X \to \mathrm{R}$ defined on a subset X of a normed space E admits extensions over E that preserves any subadditive modulus ω of f. The least and the greatest of such extensions are respectively:

$$f_*(x) := \sup_{y \in X} \{f(y) - \omega(|x-y|)\},$$

$$f^*(x) := \inf_{y \in X} \{f(y) + \omega(|x-y|)\}.$$

As remarked, any subadditive modulus of continuity is uniformly continuous: in fact, it admits itself as a modulus of continuity. Therefore, f_* and f^* are respectively inferior and superior envelopes of ω-continuous families; hence still ω-continuous. Incidentally, by the Kuratowski embedding any metric space is isometric to a subset of a normed space. Hence, special uniformly continuous real-valued functions are essentially the restrictions of uniformly continuous functions on normed spaces. In particular, this construction provides a quick proof of the Tietze extension theorem on compact metric spaces. However, for mappings with values in more general Banach spaces than R, the situation is quite more complicated; the first non-trivial result in this direction is the Kirszbraun theorem.

Concave Moduli, and Lipschitz Approximation

Every special uniformly continuous real-valued function $f: X \to \mathrm{R}$ defined on the metric space X is uniformly approximable by means of Lipschitz functions. Moreover, the speed of convergence in terms of the Lipschitz constants of the approximations is strictly related to the modulus of continuity of f.

Precisely, let ω be the minimal concave modulus of continuity of f, which is

$$\omega(t) = \inf\left\{at + b : \forall x \in X, \forall x' \in X \mid f(x) - f(x') \mid \leq a \mid x - x' \mid + b\right\}.$$

Let $\delta(s)$ be the uniform distance between the function f and the set Lip_s of all Lipschitz real-valued functions on C having Lipschitz constant s :

$$\delta(s) := \inf\left\{\| f - u \|_{\infty,X} : u \in \text{Lip}_s\right\} \leq +\infty.$$

Then the functions $\omega(t)$ and $\delta(s)$ can be related with each other via a Legendre transformation: more precisely, the functions $2\delta(s)$ and $-\omega(-t)$ (suitably extended to $+\infty$ outside their domains of finiteness) are a pair of conjugated convex functions, for

$$2\delta(s) = \sup_{t\geq 0}\left\{\omega(t) - st\right\},$$

$$\omega(t) = \inf_{s\geq 0}\left\{2\delta(s) + st\right\}.$$

Since $\omega(t) = o(1)$ for $t \to 0^+$, it follows that $\delta(s) = o(1)$ for $s \to +\infty$, that exactly means that f is uniformly approximable by Lipschitz functions. Correspondingly, an optimal approximation is given by the functions

$$f_s := \delta(s) + \inf_{y\in X}\{f(y) + sd(x,y)\}, \quad \text{for } s \in \text{dom}(\delta):$$

each function f_s has Lipschitz constant s and

$$\| f - f_s \|_{\infty,X} = \delta(s);$$

in fact, it is the greatest s-Lipschitz function that realize the distance $\delta(s)$.

For example, the α-Hölder real-valued functions on a metric space are characterized as those functions that can be uniformly approximated by s-Lipschitz functions with speed of convergence $O(s^{-\frac{\alpha}{1-\alpha}})$, while the almost Lipschitz functions are characterized by an exponential speed of convergence $O(e^{-as})$.

Examples of Use

- Let $f: [a, b] \to \mathrm{R}$ a continuous function. In the proof that f is Riemann integrable, one usually bounds the distance between the upper and lower Riemann sums with respect to the Riemann partition $P := \{t_0, ..., t_n\}$ in terms of the modulus of continuity of f and the modulus of the partition P:

$$S^*(f;P) - S_*(f;P) \leq (b-a)\omega(|P|).$$

History

Steffens (2006, p. 160) attributes the first usage of omega for the modulus of continuity to Lebesgue (1909, p. 309/p. 75) where omega refers to the oscillation of a Fourier transform. De la Vallée Poussin (1919, pp. 7-8) mentions both names (1) "modulus of continuity" and (2) "modulus of oscillation" and then concludes "but we choose (1) to draw attention to the usage we will make of it".

The Translation Group of L^p Functions, and Moduli of Continuity L^p.

Let $1 \leq p$; let $f: \mathrm{R}^n \to \mathrm{R}$ a function of class L^p, and let $h \in \mathrm{R}^n$. The h-translation of f, the function defined by $(\tau_h f)(x) := f(x-h)$, belongs to the L^p class; moreover, if $1 \leq p < \infty$, then as $||h|| \to 0$ we have:

$$|| \tau_h f - f ||_p = o(1).$$

Therefore, since translations are in fact linear isometries, also

$$|| \tau_{v+h} f - \tau_v f ||_p = o(1),$$

as $||h|| \to 0$, uniformly on $v \in \mathrm{R}^n$.

In other words, the map $h \to \tau_h$ defines a strongly continuous group of linear isometries of L^p. In the case $p = \infty$ the above property does not hold in general: actually, it exactly reduces to the uniform continuity, and defines the uniform continuous functions. This leads to the following definition, that generalizes the notion of a modulus of continuity of the uniformly continuous functions: a modulus of continuity L^p for a measurable function

$f: X \to R$ is a modulus of continuity $\omega: [0, \infty] \to [0, \infty]$ such that

$$\| \tau_h f - f \|_p \leq \omega(h).$$

This way, moduli of continuity also give a quantitative account of the continuity property shared by all L^p functions.

Modulus of Continuity of Higher Orders

It can be seen that formal definition of the modulus uses notion of finite difference of first order:

$$\omega_f(\delta) = \omega(f,\delta) = \sup_{x;|h|<\delta;} |\Delta_h(f,x)|.$$

If we replace that difference with a difference of order n we get a modulus of continuity of order n:

$$\omega_n(f,\delta) = \sup_{x;|h|<\delta;} |\Delta_h^n(f,x)|.$$

Lipschitz Continuity

In mathematical analysis, Lipschitz continuity, named after Rudolf Lipschitz, is a strong form of uniform continuity for functions. Intuitively, a Lipschitz continuous function is limited in how fast it can change: for every pair of points on the graph of this function, the absolute value of the slope of the line connecting them is no greater than a definite real number; this bound is called the function's "Lipschitz constant" (or "modulus of uniform continuity").

In the theory of differential equations, Lipschitz continuity is the central condition of the Picard–Lindelöf theorem which guarantees the existence and uniqueness of the solution to an initial value problem. A special type of Lipschitz continuity, called contraction, is used in the Banach fixed point theorem.

The concept of Lipschitz continuity is well-defined on metric spaces. A generalization of Lipschitz continuity is called Hölder continuity.

Definitions

Given two metric spaces (X, d_X) and (Y, d_Y), where d_X denotes the metric on the set X and d_Y is the metric on set Y (for example, Y might be the set of real numbers R with the metric $d_Y(x, y) = |x - y|$, and X might be a subset of R), a function $f : X \to Y$ is called Lipschitz continuous if there exists a real constant $K \geq 0$ such that, for all x_1 and x_2 in X,

$$d_Y(f(x_1), f(x_2)) \leq K d_X(x_1, x_2).$$

Any such K is referred to as a Lipschitz constant for the function *f*. The smallest constant is sometimes called the (best) Lipschitz constant; however in most cases the latter notion is less relevant. If $K = 1$ the function is called a short map, and if $0 \leq K < 1$ the function is called a contraction.

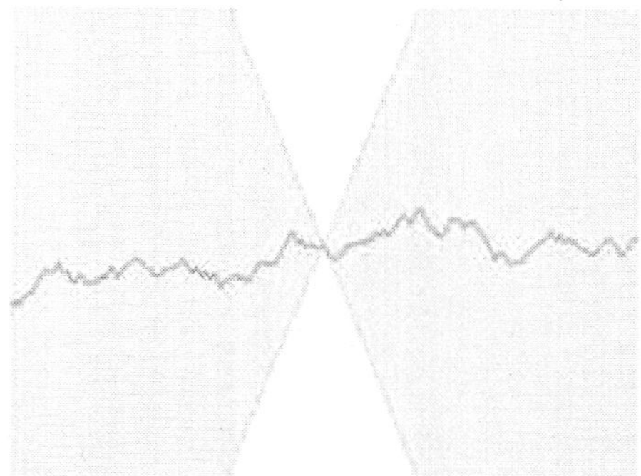

Figure: *For a Lipschitz continuous function, there is a double cone (shown in white) whose vertex can be translated along the graph, so that the graph always remains entirely outside the cone.*

The inequality is (trivially) satisfied if $x_1 = x_2$. Otherwise, one can equivalently define a function to be Lipschitz continuous if and only if there exists a constant $K \geq 0$ such that, for all $x_1 \neq x_2$,

$$\frac{d_Y(f(x_1), f(x_2))}{d_X(x_1, x_2)} \leq K.$$

For real-valued functions of several real variables, this holds if and only if the absolute value of the slopes of all secant lines are bounded by K. The set of lines of slope K passing through a point on the graph of the function forms a circular cone, and a function is Lipschitz if and only if the graph of the function everywhere lies completely outside of this cone.

A function is called locally Lipschitz continuous if for every x in X there exists a neighbourhood U of x such that f restricted to U is Lipschitz continuous. Equivalently, if X is a locally compact metric space, then f is locally Lipschitz if and only if it is Lipschitz continuous on every compact subset of X. In spaces that are not locally compact, this is a necessary but not a sufficient condition.

More generally, a function f defined on X is said to be Hölder continuous or to satisfy a Hölder condition of order $\alpha > 0$ on X if there exists a constant $M > 0$ such that

$$d_Y(f(x), f(y)) \leq M d_X(x, y)^\alpha$$

for all x and y in X. Sometimes a Hölder condition of order α is also called a uniform Lipschitz condition of order $\alpha > 0$.

If there exists a $K \geq 1$ with

$$\frac{1}{K} d_X(x_1, x_2) \leq d_Y(f(x_1), f(x_2)) \leq K d_X(x_1, x_2)$$

then f is called bilipschitz (also written bi-Lipschitz). A bilipschitz mapping is injective, and is in fact a homeomorphism onto its image. A bilipschitz function is the same thing as an injective Lipschitz function whose inverse function is also Lipschitz. Surjective bilipschitz functions are exactly the isomorphisms of metric spaces.

Examples

Lipschitz Continuous Functions The function $f(x) = \Phi x^2 + 5$ defined for all real numbers is Lipschitz continuous with the Lipschitz constant $K = 1$, because it is everywhere differentiable and the absolute value of the derivative is bounded above by 1.

Likewise, the sine function is Lipschitz continuous because its derivative, the cosine function, is bounded above by 1 in absolute value.

The function $f(x) = |x|$ defined on the reals is Lipschitz continuous with the Lipschitz constant equal to 1, by the reverse triangle inequality. This is an example of a Lipschitz continuous function that is not differentiable. More generally, a norm on a vector space is Lipschitz continuous with respect to the associated metric, with the Lipschitz constant equal to 1.

Continuous functions that are not (globally) Lipschitz continuous: The function $f(x) = \Phi x$ defined on [0, 1] is not

Lipschitz continuous. This function becomes infinitely steep as x approaches 0 since its derivative becomes infinite. However, it is uniformly continuous as well as Hölder continuous of class $C^{0,\alpha}$ for $\alpha \leq 1/2$.

Differentiable functions that are not (globally) Lipschitz continuous: The function $f(x) = x^{3/2}\sin(1/x)$ where $x \neq 0$ and $f(0) = 0$, restricted on [0, 1], gives an example of a function that is differentiable on a compact set while not locally Lipschitz because its derivative function is not bounded.

Analytic functions that are not (globally) Lipschitz continuous: The exponential function becomes arbitrarily steep as $x \rightarrow \infty$, and therefore is not globally Lipschitz continuous, despite being an analytic function.

The function $f(x) = x^2$ with domain all real numbers is not Lipschitz continuous. This function becomes arbitrarily steep as x approaches infinity. It is however locally Lipschitz continuous.

Properties

An Everywhere Differentiable Function g: $R \rightarrow R$ is Lipschitz continuous (with $K = \sup |g2\ (x)|$) if and only if it has bounded first derivative; one direction follows from the mean value theorem. In particular, any continuously differentiable function is locally Lipschitz, as continuous functions are locally bounded so its gradient is locally bounded as well.

A Lipschitz Function $g : R \rightarrow R$ is absolutely continuous and therefore is differentiable almost everywhere, that is, differentiable at every point outside a set of Lebesgue measure zero. Its derivative is essentially bounded in magnitude by the Lipschitz constant, and for $a < b$, the difference $g(b) - g(a)$ is equal to the integral of the derivative $g2$ on the interval $[a, b]$.

Conversely, if $f : I \rightarrow R$ is absolutely continuous and thus differentiable almost everywhere, and satisfies $|f2\ (x)| \leq K$ for almost all x in I, then f is Lipschitz continuous with Lipschitz constant at most K.

More generally, Rademacher's theorem extends the differentiability result to Lipschitz mappings between Euclidean spaces: a Lipschitz map $f : U \rightarrow R^m$, where U is an open set in

R^n, is almost everywhere differentiable. Moreover, if K is the best Lipschitz constant of f, then $\| Df(x) \| \leq K$ whenever the total derivative Df exists.

For a Differentiable Lipschitz Map $f: U \to R^m$ the inequality $\| Df \|_{\infty,U} \leq K$ holds for the best Lipschitz constant of f, and it turns out to be an equality if the domain U is convex.

Suppose that $\{f_n\}$ is a sequence of Lipschitz continuous mappings between two metric spaces, and that all f_n have Lipschitz constant bounded by some K. If f_n converges to a mapping f uniformly, then f is also Lipschitz, with Lipschitz constant bounded by the same K. In particular, this implies that the set of real-valued functions on a compact metric space with a particular bound for the Lipschitz constant is a closed and convex subset of the Banach space of continuous functions. This result does not hold for sequences in which the functions may have unbounded Lipschitz constants, however. In fact, the space of all Lipschitz functions on a compact metric space is dense in the Banach space of continuous functions, an elementary consequence of the Stone–Weierstrass theorem.

Every Lipschitz continuous map is uniformly continuous, and hence a fortiori continuous. More generally, a set of functions with bounded Lipschitz constant forms an equicontinuous set. The Arzelà–Ascoli theorem implies that if $\{f_n\}$ is a uniformly bounded sequence of functions with bounded Lipschitz constant, then it has a convergent subsequence. By the result of the previous paragraph, the limit function is also Lipschitz, with the same bound for the Lipschitz constant. In particular the set of all real-valued Lipschitz functions on a compact metric space X having Lipschitz constant $\leq K$ is a locally compact convex subset of the Banach space $C(X)$.

For a family of Lipschitz continuous functions f_α with common constant, the function $\sup_\alpha f_\alpha$ (and $\inf_\alpha f_\alpha$) is Lipschitz continuous as well, with the same Lipschitz constant, provided it assumes a finite value at least at a point.

If U is a Subset of The Metric Space M and $f: U \to R$ is a Lipschitz continuous function, there always exist Lipschitz continuous maps $M \to R$ which extend f and have the same Lipschitz constant as f . An extension is provided by

$$\tilde{f}(x) := \inf_{u \in U}\{f(u) + kd(x,u)\},$$

where k is a Lipschitz constant for f on U.

Lipschitz Manifolds

Let U and V be two open sets in R^n. A function $T: U \to V$ is called bi-Lipschitz if it is a Lipschitz homeomorphism onto its image, and its inverse is also Lipschitz.

Using bi-Lipschitz mappings, it is possible to define a Lipschitz structure on a topological manifold, since there is a pseudogroup structure on bi-Lipschitz homeomorphisms. This structure is intermediate between that of a piecewise-linear manifold and a smooth manifold. In fact a PL structure gives rise to a unique Lipschitz structure; it can in that sense 'nearly' be smoothed.

One-sided Lipschitz

Let $F(x)$ be an upper semi-continuous function of x, and that $F(x)$ is a closed, convex set for all x. Then F is one-sided Lipschitz if

$$(x_1 - x_2)^T (F(x_1) - F(x_2)) \le C \| x_1 - x_2 \|^2$$

for some C for all x_1 and x_2.

It is possible that the function F could have a very large Lipschitz constant but a moderately sized, or even negative, one-sided Lipschitz constant. For example the function

$$\begin{cases} F : \mathrm{R}^2 \to \mathrm{R}, \\ F(x,y) = -50(y - cos(x)) \end{cases}$$

has Lipschitz constant K = 50 and a one-sided Lipschitz constant C = 0.

Equicontinuity

In mathematical analysis, a family of functions is equicontinuous if all the functions are continuous and they have equal variation over a given neighbourhood, in a precise sense described herein. In particular, the concept applies to countable families, and thus *sequences* of functions.

The equicontinuity appears in the formulation of Ascoli's theorem, which states that a subset of $C(X)$, the space of continuous functions on a compact Hausdorff space X, is compact if and only if it is closed, pointwise bounded and equicontinuous. As a corollary, a sequence in $C(X)$ is uniformly convergent if and only if it is equicontinuous and converges pointwise to a function (not necessarily continuous a-priori). In particular, the limit of an equicontinuous pointwise convergent sequence of continuous functions f_n on either metric space or locally compact space is continuous. If, in addition, f_n are holomorphic, then the limit is also holomorphic.

The uniform boundedness principle states that a pointwise bounded family of continuous linear operators between Banach spaces is equicontinuous.

Definition

Let X and Y be two metric spaces, and F a family of functions from X to Y.

The family F is equicontinuous at a point $x_0 \in X$ if for every $\varepsilon > 0$, there exists a $\delta > 0$ such that $d(f(x_0), f(x)) < \varepsilon$ for all $f \in F$ and all x such that $d(x_0, x) < \delta$. The family is equicontinuous if it is equicontinuous at each point of X.

The family F is uniformly equicontinuous if for every $\varepsilon > 0$, there exists a $\delta > 0$ such that $d(f(x_1), f(x_2)) < \varepsilon$ for all $f \in F$ and all $x_1, x_2 \in X$ such that $d(x_1, x_2) < \delta$.

For comparison, the statement 'all functions f in F are continuous' means that for every $\varepsilon > 0$, every $f \in F$, and every $x_0 \in X$, there exists a $\delta > 0$ such that $d(f(x_0), f(x)) < \varepsilon$ for all $x \in X$ such that $d(x_0, x) < \delta$.

- For *continuity*, δ may depend on ε, x_0 and f.
- For *uniform continuity*, δ may depend on ε, and f.
- For *equicontinuity*, δ may depend on ε, and x_0.
- For *uniform equicontinuity*, δ may solely depend on ε.

More generally, when X is a topological space, a set F of functions from X to Y is said to be equicontinuous at x if for every $\varepsilon > 0$, x has a neighbourhood U_x such that

$$d_Y(f(y), f(x)) < \epsilon$$

for all $y \in U_x$ and $f \in F$. This definition usually appears in the context of topological vector spaces.

When X is compact, a set is uniformly equicontinuous if and only if it is equicontinuous at every point, for essentially the same reason as that uniform continuity and continuity coincide on compact spaces.

Some basic properties follow immediately from the definition. Every finite set of continuous functions is equicontinuous. The closure of an equicontinuous set is again equicontinuous. Every member of a uniformly equicontinuous set of functions is uniformly continuous, and every finite set of uniformly continuous functions is uniformly equicontinuous.

Examples

- A set of functions with the same Lipschitz constant is (uniformly) equicontinuous. In particular, this is the case if the set consists of functions with derivatives bounded by the same constant.
- Uniform boundedness principle gives a sufficient condition for a set of continuous linear operators to be equicontinuous.
- A family of iterates of an analytic function is equicontinuous on the Fatou set.

Non Examples

- The sequence of functions $f_n(x) = \arctan(nx)$, is not equicontinuous because the definition is violated at $x_0=0$

Equicontinuity and Uniform Convergence

Let X be a compact Hausdorff space, and equip $C(X)$ with the uniform norm, thus making $C(X)$ a Banach space, hence a metric space. Then Ascoli's theorem states that a subset of $C(X)$ is compact if and only if it is closed, pointwise bounded and equicontinuous. This is analogous to the Heine-Borel theorem, which states that subsets of R^n are compact if and only if they are closed and bounded. As a corollary, every uniformly bounded equicontinuous sequence in $C(X)$ contains a subsequence that converges uniformly to a continuous function on X.

In view of Ascoli's theorem, a sequence in $C(X)$ converges uniformly if and only if it is equicontinuous and converges pointwise. The hypothesis of the statement can be weakened a bit: a sequence in $C(X)$ converges uniformly if it is equicontinuous and converges pointwise on a dense subset to some function on X (not assumed continuous). This weaker version is typically used to prove Ascoli's theorem for separable compact spaces. Another consequence is that the limit of an equicontinuous pointwise convergent sequence of continuous functions on a metric space, or on a locally compact space, is continuous. In the above, the hypothesis of compactness of X cannot be relaxed. To see that, consider a compactly supported continuous function g on R with $g(0) = 1$, and consider the equicontinuous sequence of functions $\{f_n\}$ on R defined by $f_n(x) = g(x - n)$. Then, f_n converges pointwise to 0 but does not converge uniformly to 0.

This criterion for uniform convergence is often useful in real and complex analysis. Suppose we are given a sequence of continuous functions that converges pointwise on some open subset G of R^n. As noted above, it actually converges uniformly on a compact subset of G if it is equicontinuous on the compact set. In practice, showing the equicontinuity is often not so difficult. For example, if the sequence consists of differentiable functions or functions with some regularity (e.g., the functions are solutions of a differential equation), then the mean value theorem or some other kinds of estimates can be used to show the sequence is equicontinuous. It then follows that the limit of the sequence is continuous on every compact subset of G; thus, continuous on G. A similar argument can be made when the functions are holomorphic. One can use, for instance, Cauchy's estimate to show the equicontinuity (on a compact subset) and conclude that the limit is holomorphic. Note that the equicontinuity is essential here. For example, $f_n(x) = \arctan n\,x$ converges to a multiple of the discontinuous sign function.

Generalizations

Equicontinuity Families of Linear Operators: Let E, F be Banach spaces, and Γ be a family of continuous linear operators from E into F. Then Γ is equicontinuous if and only if

$$\sup\{\| T \| : T \in \Gamma\} < \infty$$

that is, Γ is uniformly bounded in operator norm. Also, by linearity, Γ is uniformly equicontinuous if and only if it is equicontinuous at 0.

The uniform boundedness principle (also known as the Banach–Steinhaus theorem) states that Γ is equicontinuous if it is pointwise bounded; i.e., $\sup \{\| T(x) \|: T \in \Gamma\} < \infty$ for each $x \in E$. The result can be generalized to a case when F is locally convex and E is a barreled space.

Alaoglu's theorem states that if E is a topological vector space, then every equicontinuous subset of E^* is weak-* relatively compact.

Equicontinuity in Topological Spaces

The most general scenario in which equicontinuity can be defined is for topological spaces whereas *uniform* equicontinuity requires the filter of neighbourhoods of one point to be somehow comparable with the filter of neighbourhood of another point. The latter is most generally done via a uniform structure, giving a uniform space. Appropriate definitions in these cases are as follows:

A set A of functions continuous between two topological spaces X and Y is topologically equicontinuous at the points $x \in X$ and $y \in Y$ if for any open set O about y, there are neighbourhoods U of x and V of y such that for every $f \in A$, if the intersection of $f[U]$ and V is nonempty, $f(U) \subseteq O$. One says A is said to be topologically equicontinuous at $x \in X$ if it is topologically equicontinuous at x and y for each $y \in Y$. Finally, A is equicontinuous if it is equicontinuous at x for all points $x \in X$.

A set A of continuous functions between two uniform spaces X and Y is uniformly equicontinuous if for every element W of the uniformity on Y, the set

$$\{ (u,v) \in X \times X: \text{for all } f \in A.\ (f(u),f(v)) \in W \}$$

is a member of the uniformity on X

A Weaker Concept is That of Even Continuity: A set A of continuous functions between two topological spaces X and Y is said to be evenly continuous at $x \in X$ and $y \in Y$ if given any open set O containing y there are neighbourhoods U of x and V of y such that $f[U] \subseteq O$ whenever $f(x) \in V$. It is evenly continuous at x if it is evenly continuous at x and y for every $y \in Y$, and evenly continuous if it is evenly continuous at x for every $x \in X$.

For metric spaces, there are standard topologies and uniform structures derived from the metrics, and then these general definitions are equivalent to the metric-space definitions.

Chapter 2

Stochastic Equicontinuity

In estimation theory in statistics, stochastic equicontinuity is a property of estimators or of estimation procedures that is useful in dealing with their asymptotic behviour as the amount of data increases. It is a version of equicontinuity used in the context of functions of random variables: that is, random functions. The property relates to the rate of convergence of sequences of random variables and requires that this rate is essentially the same within a region of the parameter space being considered.

For instance, stochastic equicontinuity, along with other conditions, can be used to show uniform weak convergence, which can be used to prove the convergence of extremum estimators.

Absolute Continuity

In calculus, absolute continuity is a smoothness property of functions that is stronger than continuity and uniform continuity. The notion of absolute continuity allows one to obtain generalisations of the relationship between the two central operations of calculus, differentiation and integration, expressed by the fundamental theorem of calculus in the framework of Riemann integration. Such generalisations are often formulated in terms of Lebesgue integration. For real-valued functions on the real line two interrelated notions appear, *absolute continuity of functions* and *absolute continuity of measures.* These two notions are generalized in different directions. The usual derivative of a function is related to the *Radon–Nikodym derivative*, or *density*, of a measure.

Absolute Continuity of Functions

It may happen that a continuous function f is differentiable almost everywhere on [0,1], its derivative f' is Lebesgue integrable, and nevertheless the integral of f' differs from the increment of f. For example, this happens for the Cantor function, which means that this function is not absolutely continuous.

Definition

Let I be an interval in the real line R. A function $f: I \to$ R is absolutely continuous on I if for every positive number ϵ, there is a positive number δsuch that whenever a finite sequence of pairwise disjoint sub-intervals (x_k, y_k) of I satisfies

$$\sum_k |y_k - x_k| < \delta$$

then

$$\sum_k |f(y_k) - f(x_k)| < \epsilon.$$

The collection of all absolutely continuous functions on I is denoted AC(I).

Equivalent Definitions

The following conditions on a real-valued function f on a compact interval $[a,b]$ are equivalent:

(1) f is absolutely continuous;

(2) f has a derivative f' almost everywhere, the derivative is Lebesgue integrable, and

$$f(x) = f(a) + \int_a^x f'(t)dt$$

for all x on $[a,b]$;

(3) there exists a Lebesgue integrable function g on $[a,b]$ such that

$$f(x) = f(a) + \int_a^x g(t)dt$$

for all x on $[a,b]$.

If these equivalent conditions are satisfied then necessarily $g = f'$ almost everywhere.

Equivalence between (1) and (3) is known as the fundamental theorem of Lebesgue integral calculus, due to Lebesgue.

Properties

- The sum and difference of two absolutely continuous functions are also absolutely continuous. If the two functions are defined on a bounded closed interval, then their product is also absolutely continuous.
- If an absolutely continuous function is defined on a bounded closed interval and is nowhere zero then its reciprocal is absolutely continuous.
- Every absolutely continuous function is uniformly continuous and, therefore, continuous. Every Lipschitz-continuous function is absolutely continuous.
- If f: $[a,b] \to$ R is absolutely continuous, then it is of bounded variation on $[a,b]$.
- If f: $[a,b] \to$ R is absolutely continuous, then it has the Luzin N property (that is, for any $L \subseteq [a,b]$ such that $\lambda(L) = 0$, it holds that $\lambda(f(L)) = 0$, where λ stands for the Lebesgue measure on R).
- f: $I \to$ R is absolutely continuous if and only if it is continuous, is of bounded variation and has the Luzin N property.

Examples

The following functions are continuous everywhere but not absolutely continuous:

- the Cantor function;
- the function

$$f(x) = \begin{cases} 0, & \text{if } x = 0 \\ x\sin(1/x), & \text{if } x \neq 0 \end{cases}$$

on a finite interval containing the origin; the function $f(x) = x^2$ on an unbounded interval.

Generalizations

Let (X, d) be a metric space and let I be an interval in the real line R. A function f: $I \to X$ is absolutely continuous on I

if for every positive number ϵ, there is a positive number δsuch that whenever a finite sequence of pairwise disjoint sub-intervals $[x_k, y_k]$ of I satisfies

$$\sum_k |y_k - x_k| < \delta$$

then

$$\sum_k d\big(f(y_k), f(x_k)\big) < \epsilon.$$

The collection of all absolutely continuous functions from I into X is denoted AC(I; X).

A further generalization is the space AC^p(I; X) of curves f: $I \to X$ such that

$$d\big(f(s), f(t)\big) \leq \int_s^t m(\tau)\mathrm{d}\tau \text{ for all } [s,t] \subseteq I$$

for some m in the L^p space L^p(I).

Properties of These Generalizations

- Every absolutely continuous function is uniformly continuous and, therefore, continuous. Every Lipschitz-continuous function is absolutely continuous.
- If f: $[a,b] \to X$ is absolutely continuous, then it is of bounded variation on $[a,b]$.
- For $f \in$ AC^p(I; X), the metric derivative of f exists for λ-almost all times in I, and the metric derivative is the smallest $m \in L^p(I;\ \mathrm{R})$ such that

$$d\big(f(s), f(t)\big) \leq \int_s^t m(\tau)\mathrm{d}\tau \text{ for all } [s,t] \subseteq I.$$

Absolute Continuity of Measures

Definition

A measure μ on Borel subsets of the real line is absolutely continuous with respect to Lebesgue measure λ (in other words, dominated by λ) if $\mu(A) = 0$ for every set A for which $\lambda(A) = 0$. This is written as "$\mu << \lambda$".

In most applications, if a measure on the real line is simply said to be absolutely continuous — without specifying with respect to which other measure it is absolutely continuous —

then absolute continuity with respect to Lebesgue measure is meant.

The same holds for R^n for all n=1,2,3,...

Equivalent Definitions

The following conditions on a finite measure μ on Borel subsets of the real line are equivalent:

(1) μ is absolutely continuous;

(2) for every positive number *e* there is a positive number δ such that $\mu(A) < \varepsilon$ for all Borel sets A of Lebesgue measure less than *d*;

(3) there exists a Lebesgue integrable function g on the real line such that

$$\mu(A) = \int_A g d\lambda$$

for all Borel subsets A of the real line.

Any other function satisfying (3) is equal to g almost everywhere. Such a function is called Radon-Nikodym derivative, or density, of the absolutely continuous measure μ.

Equivalence between (1), (2) and (3) holds also in R^n for all n=1,2,3,...

Thus, the absolutely continuous measures on R^n are precisely those that have densities; as a special case, the absolutely continuous probability measures are precisely the ones that have probability density functions.

Generalizations

If μ and ν are two measures on the same measurable space then μ is said to be absolutely continuous with respect to ν, or dominated by ν if $\mu(A) = 0$ for every set A for which $\nu(A) = 0$. This is written as "$\mu \ll \nu$". In symbols:

$$\mu \ll \nu \Leftrightarrow \left(\nu(A) = 0 \Rightarrow \mu(A) = 0\right).$$

Absolute continuity of measures is reflexive and transitive, but is not antisymmetric, so it is a preorder rather than a partial order. Instead, if $\mu \ll \nu$ and $\nu \ll \mu$, the measures μ and ν are said to be equivalent. Thus absolute continuity induces a partial ordering of such equivalence classes.

If μ is a signed or complex measure, it is said that μ is absolutely continuous with respect to ν if its variation $|\mu|$ satisfies $|\mu| \ll \nu$; equivalently, if every set A for which $\nu(A) = 0$ is μ-null.

The Radon–Nikodym theorem states that if μ is absolutely continuous with respect to ν, and both measures are σ-finite, then μ has a density, or "Radon-Nikodym derivative", with respect to ν, which means that there exists a ν-measurable function f taking values in $[0, +\infty)$, denoted by $f = {}^{d\mu}/_{d\nu}$, such that for any ν-measurable set A we have

$$\mu(A) = \int_A f \, d\nu.$$

Singular Measures

Via Lebesgue's decomposition theorem, every measure can be decomposed into the sum of an absolutely continuous measure and a singular measure.

Relation between the two Notions of Absolute Continuity

A finite measure μ on Borel subsets of the real line is absolutely continuous with respect to Lebesgue measure if and only if the point function

$$F(x) = \mu((-\infty, x])$$

is locally an absolutely continuous real function. In other words, a function is locally absolutely continuous if and only if its distributional derivative is a measure that is absolutely continuous with respect to the Lebesgue measure.

If the absolute continuity holds then the Radon-Nikodym derivative of μ is equal almost everywhere to the derivative of F.

More generally, the measure μ is assumed to be locally finite (rather than finite) and $F(x)$ is defined as $\mu((0,x])$ for $x>0$, 0 for $x=0$, and $-\mu((x,0])$ for $x<0$. In this case μ is the Lebesgue-Stieltjes measure generated by F. The relation between the two notions of absolute continuity still holds.

Distribution (Mathematics)

In mathematical analysis, distributions (or generalized functions) are objects that generalize functions. Distributions make it possible to differentiate functions whose derivatives do not exist in the classical sense. In particular, any locally integrable function has a distributional derivative. Distributions are widely used to formulate generalized solutions of partial differential equations. Where a classical solution may not exist or be very difficult to establish, a distribution solution to a differential equation is often much easier. Distributions are also important in physics and engineering where many problems naturally lead to differential equations whose solutions or initial conditions are distributions, such as the Dirac delta function (which is historically called a "function" even though it is not considered a proper function mathematically).

Generalized functions were introduced by Sergei Sobolev in 1935. They were re-introduced in the late 1940s by Laurent Schwartz, who developed a comprehensive theory of distributions.

Basic idea

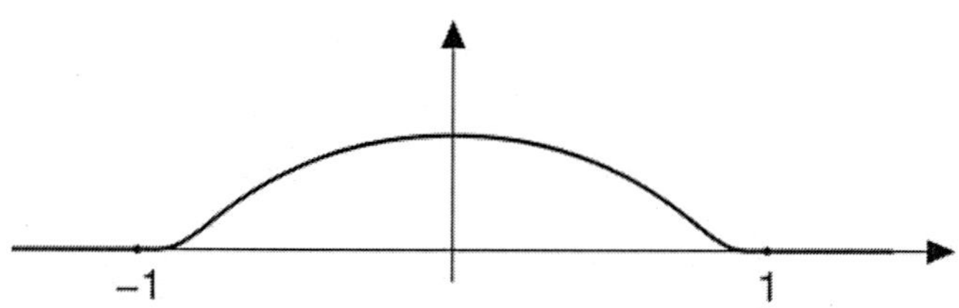

Figure: *A typical test function, the bump function* ψ (x)*. It is smooth (infinitely differentiable) and has compact support (is zero outside an interval, in this case the interval [–1, 1]).*

Distributions are a class of linear functionals that map a set of *test functions* (conventional and well-behaved functions) onto the set of real numbers. In the simplest case, the set of test functions considered is D(R), which is the set of functions $\varphi : R \to R$ having two properties:

- φ is smooth (infinitely differentiable);
- φ has compact support (is identically zero outside some bounded interval).

Then, a distribution d is a linear mapping $D(R) \to R$. Instead of writing $d(\varphi)$, where φ is a test function in D(R), it is conventional to write $\langle d, \varphi \rangle$. A simple example of a distribution is the Dirac delta δ, defined by

$$\delta(\varphi) = \langle \delta, \varphi \rangle = \varphi(0).$$

There are straightforward mappings from both locally integrable functions and probability distributions to corresponding distributions, as discussed below. However, not all distributions can be formed in this manner.

Suppose That f : R → R is a locally integrable function, and let φ : R → R be a test function in D(R). We can then define a corresponding distribution T_f by:

$$\langle T_f, \varphi \rangle = \int_{\mathrm{R}} f(x)\varphi(x)dx.$$

This integral is a real number which linearly and continuously depends on φ. This suggests the requirement that a distribution should be linear and continuous over the space of test functions D(R), which completes the definition. In a conventional abuse of notation, *f* may be used to represent both the original function *f* and the distribution T_f derived from it.

Similarly, if *p* is a probability distribution on the reals and φ is a test function, then a corresponding distribution T_p may be defined by:

$$\langle T_p, \varphi \rangle = \int_{\mathrm{R}} \varphi dp$$

Again, this integral continuously and linearly depends on φ, so that T_p is in fact a distribution.

Such distributions may be multiplied with real numbers and can be added together, so they form a real vector space. In general it is not possible to define a multiplication for distributions, but distributions may be multiplied with infinitely differentiable functions.

It's desirable to choose a definition for the derivative of a distribution which, at least for distributions derived from locally integrable functions, has the property that $T_f' = T_{f'}$. If φ is a test function, we can use integration by parts to see that

$$\langle f', \varphi \rangle = \int_{\mathrm{R}} f'\varphi dx = \left[f(x)\varphi(x)\right]_{-\infty}^{\infty} - \int_{\mathrm{R}} f\varphi' dx = -\langle f, \varphi' \rangle$$

where the last equality follows from the fact that φ is zero outside of a bounded set. This suggests that if *S* is a *distribution*, we should define its derivative *S*2 by

$$\langle S', \varphi \rangle = -\langle S, \varphi' \rangle.$$

It turns out that this is the proper definition; it extends the ordinary definition of derivative, every distribution becomes infinitely differentiable and the usual properties of derivatives hold.

Example: Recall that the Dirac delta (so-called Dirac delta function) is the distribution defined by

$$\langle \delta, \varphi \rangle = \varphi(0)$$

It is the derivative of the distribution corresponding to the Heaviside step function H: For any test function φ,

$$\langle H', \varphi \rangle = -\langle H, \varphi' \rangle = -\int_{-\infty}^{\infty} H(x)\varphi'(x)dx = -\int_{0}^{\infty} \varphi'(x)dx = \varphi(0) - \varphi(\infty) = \varphi(0) = \langle \delta, \varphi \rangle,$$

so $H2 = \delta$. Note, $\varphi(\infty) = 0$ because of compact support. Similarly, the derivative of the Dirac delta is the distribution

$$\langle \delta', \varphi \rangle = -\varphi'(0).$$

This latter distribution is our first example of a distribution which is derived from neither a function nor a probability distribution.

Test Functions and Distributions

In the following, real-valued distributions on an open subset U of R^n will be formally defined. With minor modifications, one can also define complex-valued distributions, and one can replace R^n by any (paracompact) smooth manifold.

The first object to define is the space $\mathrm{D}(U)$ of test functions on U. Once this is defined, it is then necessary to equip it with a topology by defining the limit of a sequence of elements of $\mathrm{D}(U)$. The space of distributions will then be given as the space of continuous linear functionals on $\mathrm{D}(U)$.

Test Function Space

The space $\mathrm{D}(U)$ of test functions on U is defined as follows. A function $\varphi : U \to \mathrm{R}$ is said to have compact support if there exists a compact subset K of U such that $\varphi(x) = 0$ for all x in $U \setminus K$. The elements of $\mathrm{D}(U)$ are the infinitely differentiable functions $\varphi : U \to \mathrm{R}$ with compact support – also known as bump functions. This is a real vector space. It can be given a topology by defining the limit of a sequence of elements of $\mathrm{D}(U)$. A sequence (φ_k) in $\mathrm{D}(U)$ is said to converge to $\varphi \in \mathrm{D}(U)$ if the following two conditions hold (Gelfand & Shilov 1966–1968, v. 1, §1.2):

- There is a compact set $K \subset U$ containing the supports of all φ_k:

  $\bigcup_k \text{supp}(\varphi_k) \subset K.$

- For each multiindex α, the sequence of partial derivatives $D^\alpha\varphi_k$ tends uniformly to $D^\alpha\varphi$.

With this definition, D(U) becomes a complete locally convex topological vector space satisfying the Heine–Borel property (Rudin 1991, §6.4–5).

This topology can be placed in the context of the following general construction: let

$X = \bigcup_i X_i$ be a countable increasing union of locally convex topological vector spaces and $\iota_i : X_i \to X$ be the inclusion maps. In this context, the inductive limit topology, or final topology, τ on X is the finest locally convex vector space topology making all the inclusion maps ι_i continuous. The topology τ can be explicitly described as follows: let β be the collection of convex balanced subsets W of X such that $W \cap X_i$ is open for all i. A base for the inductive limit topology τ then consists of the sets of the form $x + W$, where x in X and W in β.

The proof that τ is a vector space topology makes use of the assumption that each X_i is locally convex. By construction, β is a local base for τ. That any locally convex vector space topology on X must necessarily contain τ means it is the weakest one. One can also show that, for each i, the subspace topology X_i inherits from τ coincides with its original topology. When each X_i is a Fréchet space, (X, τ) is called an LF space.

Now let U be the union of U_i where $\{U_i$ is a countable$\}$ nested family of open subsets of U with compact closures K_i = U_i. Then we have the countable increasing union

$$\mathrm{D}(U) = \bigcup_i \mathrm{D}_{K_i}$$

where D_{Ki} is the set of all smooth functions on U with support lying in K_i. On each D_{Ki}, consider the topology given by the seminorms

$$\| \varphi \|_\alpha = \max_{x \in K_i} \left| \delta^\alpha \varphi \right|,$$

i.e. the topology of uniform convergence of derivatives of arbitrary order. This makes each D_{Ki} a Fréchet space. The resulting LF space structure on D(U) is the topology described in the beginning of the section.

On D(U), one can also consider the topology given by the seminorms

$$\| \varphi \|_{\alpha,K_i} = \max_{x \in K_i} \left| \delta^{\alpha} \varphi \right|.$$

However, this topology has the disadvantage of not being complete. On the other hand, because of the particular features of D_{Ki}'s, a set this bounded with respect to τ if and only if it lies in some D_{Ki}'s. The completeness of ($D(U)$, τ) then follow from that of D_{Ki}'s.

The topology τ is not metrizable by the Baire category theorem, since D(U) is the union of subspaces of the first category in D(U) (Rudin 1991, §6.9).

Distributions

A distribution on U is a linear functional S : D(U) → R (or S : D(U) → C), such that

$$\lim_{n \to \infty} S(\varphi_n) = S\left(\lim_{n \to \infty} \varphi_n\right)$$

for any convergent sequence φ_n in D(U). The space of all distributions on U is denoted by D' (U). Equivalently, the vector space D' (U) is the continuous dual space of the topological vector space D(U).

The dual pairing between a distribution S in D' (U) and a test function φ in D(U) is denoted using angle brackets thus:

$$\begin{cases} D'(U) \times D(U) \to R \\ (S, \varphi) \mapsto \langle S, \varphi \rangle. \end{cases}$$

Equipped with the weak-* topology, the space D' (U) is a locally convex topological vector space. In particular, a sequence (S_k) in D' (U) converges to a distribution S if and only if

$$\langle S_k, \varphi \rangle \to \langle S, \varphi \rangle$$

for all test functions φ. This is the case if and only if S_k converges uniformly to S on all bounded subsets of D(U). (A subset E of D(U) is bounded if there exists a compact subset K of U and numbers d_n such that every φ in E has its support in K and has its n-th derivatives bounded by d_n.)

Functions as Distributions

The function $f: U \to \mathrm{R}$ is called locally integrable if it is Lebesgue integrable over every compact subset K of U. This is a large class of functions which includes all continuous functions and all L^p functions. The topology on D(U) is defined in such a fashion that any locally integrable function f yields a continuous linear functional on D(U) – that is, an element of D' (U) – denoted here by T_f, whose value on the test function φ is given by the Lebesgue integral:

$$\langle T_f, \varphi \rangle = \int_U f\varphi dx.$$

Conventionally, one abuses notation by identifying T_f with f, provided no confusion can arise, and thus the pairing between f and φ is often written

$$\langle f, \varphi \rangle = \langle T_f, \varphi \rangle.$$

If f and g are two locally integrable functions, then the associated distributions T_f and T_g are equal to the same element of D' (U) if and only if f and g are equal almost everywhere. In a similar manner, every Radon measure μ on U defines an element of D' (U) whose value on the test function φ is $\int \varphi d\mu$. As above, it is conventional to abuse notation and write the pairing between a Radon measure μ and a test function φ as $\langle \mu, \varphi \rangle$. Conversely, as shown in a theorem by Schwartz (similar to the Riesz representation theorem), every distribution which is non-negative on non-negative functions is of this form for some (positive) Radon measure.

The test functions are themselves locally integrable, and so define distributions. As such they are dense in D' (U) with respect to the topology on D' (U) in the sense that for any distribution $S \in$ D' (U), there is a sequence $\varphi_n \in$ D(U) such that

$$\langle \varphi_n, \psi \rangle \to \langle S, \psi \rangle$$

for all $\psi \in$ D(U). This follows at once from the Hahn–Banach theorem, since by an elementary fact about weak topologies the dual of D' (U) with its weak-* topology is the space D(U) (Rudin 1991, Theorem 3.10). This can also be proven more constructively by a convolution argument.

Operations on Distributions

Many operations which are defined on smooth functions with compact support can also be defined for distributions. In general, if $T : D(U) \to D(U)$ is a linear mapping of vector spaces which is continuous with respect to the weak-* topology, then it is possible to extend T to a mapping T: D' $(U) \to$ D' (U) by passing to the limit. (This approach works for more general non-linear mappings as well, provided they are assumed to be uniformly continuous.)

In practice, however, it is more convenient to define operations on distributions by means of the transpose (or adjoint transformation) (Strichartz 1994, §2.3; Trèves 1967). If $T : D(U) \to D(U)$ is a continuous linear operator, then the transpose is an operator $T^* : D(U) \to D(U)$ such that

$$\langle T\varphi, \psi \rangle = \langle \varphi, T^* \psi \rangle$$

for all φ, $\psi \in D(U)$. If such an operator T^* exists, and is continuous, then the original operator T may be extended to distributions by defining

$$Tf(\psi) = f(T^* \psi).$$

Differentiation

If $T : D(U) \to D(U)$ is given by the partial derivative

$$T\varphi = \frac{\partial \varphi}{\partial x_k}.$$

By integration by parts, if φ and ψ are in $D(U)$, then

$$\langle T\varphi, \psi \rangle = \left\langle \frac{\partial \varphi}{\partial x_k}, \psi \right\rangle = -\left\langle \varphi, \frac{\partial \psi}{\partial x_k} \right\rangle$$

so that $T^* = -T$. This is a continuous linear transformation $D(U) \to D(U)$. So, if $S \in D'(U)$ is a distribution, then the partial derivative of S with respect to the coordinate x_k is defined by the formula

$$\left\langle \frac{\partial S}{\partial x_k}, \varphi \right\rangle = -\left\langle S, \frac{\partial \varphi}{\partial x_k} \right\rangle$$

for all test functions φ. In this way, every distribution is infinitely differentiable, and the derivative in the direction x_k is a linear operator on D' (U). In general, if $\alpha = (\alpha_1, ..., \alpha_n)$ is an arbitrary multi-index and ∂^α denotes the associated mixed partial derivative operator, the mixed partial derivative $\partial^\alpha S$ of the distribution $S \in$ D' (U) is defined by

$$\left\langle \partial^\alpha S, \varphi \right\rangle = (-1)^{|\alpha|} \left\langle S, \partial^\alpha \varphi \right\rangle \text{ for all } \varphi \in \mathrm{D}(U).$$

Differentiation of distributions is a *continuous* operator on D' (U); this is an important and desirable property that is not shared by most other notions of differentiation.

Multiplication by a Smooth Function

If $m : U \to$ R is an infinitely differentiable function and S is a distribution on U, then the product mS is defined by $(mS)(\varphi) = S(m\varphi)$ for all test functions φ. This definition coincides with the transpose transformation of

$$T_m : \varphi \mapsto m\varphi$$

for $\varphi \in$ D(U). Then, for any test function ψ

$$\langle T_m \varphi, \psi \rangle = \int_U m(x)\varphi(x)\psi(x)dx = \langle \varphi, T_m \psi \rangle$$

so that $T^*_m = T_m$. Multiplication of a distribution S by the smooth function m is therefore defined by

$$mS(\psi) = \langle mS, \psi \rangle = \langle S, m\psi \rangle = S(m\psi).$$

Under multiplication by smooth functions, D' (U) is a module over the ring $C^\infty(U)$. With this definition of multiplication by a smooth function, the ordinary product rule of calculus remains valid. However, a number of unusual identities also arise. For example, the Dirac delta distribution δ is defined on R by $\langle \delta, \varphi \rangle = \varphi(0)$, so that $m\delta = m(0)\delta$. Its derivative is given by $\langle \delta 2, \varphi \rangle = -\langle \delta, \varphi 2 \rangle = -\varphi 2\ (0)$. But the product m$\delta$2 of m and δ2 is the distribution

$$m\delta' = m(0)\delta' - m'\delta = m(0)\delta' - m'(0)\delta.$$

This definition of multiplication also makes it possible to define the operation of a linear differential operator with smooth coefficients on a distribution. A linear differential operator takes a distribution $S \in$ D' (U) to another distribution given by a sum of the form

$$PS = \sum_{|\alpha| \le k} p_\alpha \partial^\alpha S$$

where the coefficients p_α are smooth functions in U. If P is a given differential operator, then the minimum integer k for which such an expansion holds for every distribution S is called the order of P. The transpose of P is given by

$$\left\langle \sum_\alpha p_\alpha \partial^\alpha S, \varphi \right\rangle = \left\langle S, \sum_\alpha (-1)^{|\alpha|} \partial^\alpha (p_\alpha \varphi) \right\rangle.$$

The space D' (U) is a D-module with respect to the action of the ring of linear differential operators.

Composition with a Smooth Function

Let S be a distribution on an open set $U \subset \mathrm{R}^n$. Let V be an open set in R^n, and $F : V \to U$. Then provided F is a submersion, it is possible to define

$$S^\circ F \in \mathrm{D}'(V).$$

This is the composition of the distribution S with F, and is also called the pullback of S along F, sometimes written

$$F^\sharp : S \mapsto F^\sharp S = S^\circ F.$$

The pullback is often denoted F^*, but this notation risks confusion with the above use of '*' to denote the transpose of a linear mapping.

The condition that F be a submersion is equivalent to the requirement that the Jacobian derivative $dF(x)$ of F is a surjective linear map for every $x \in V$. A necessary (but not sufficient) condition for extending $F^\#$ to distributions is that F be an open mapping (Hörmander 1983, Theorem 6.1.1). The inverse function theorem ensures that a submersion satisfies this condition.

If F is a submersion, then $F^\#$ is defined on distributions by finding the transpose map. Uniqueness of this extension is guaranteed since $F^\#$ is a continuous linear operator on D(U). Existence, however, requires using the change of variables formula, the inverse function theorem (locally) and a partition of unity argument.

In the special case when F is a diffeomorphism from an open subset V of R^n onto an open subset U of R^n change of variables under the integral gives

$$\int_V \varphi^\circ F(x)\psi(x)dx = \int_U \varphi(x)\psi\left(F^{-1}(x)\right)\left|\det dF^{-1}(x)\right|dx.$$

In this particular case, then, $F^{\#}$ is defined by the transpose formula:

$$\left\langle F^{\sharp}S,\varphi\right\rangle = \left\langle S,\left|\det d(F^{-1})\right|\varphi\circ F^{-1}\right\rangle.$$

Localization of Distributions

There is no way to define the value of a distribution in D' (U) at a particular point of U. However, as is the case with functions, distributions on U restrict to give distributions on open subsets of U. Furthermore, distributions are *locally determined* in the sense that a distribution on all of U can be assembled from a distribution on an open cover of U satisfying some compatibility conditions on the overlap. Such a structure is known as a sheaf.

Restriction

Let U and V be open subsets of R^n with $V \subset U$. Let $E_{VU} : D(V) \to D(U)$ be the operator which *extends by zero* a given smooth function compactly supported in V to a smooth function compactly supported in the larger set U. Then the restriction mapping ρ_{VU} is defined to be the transpose of E_{VU}. Thus for any distribution $S \in D'$ (U), the restriction $\rho_{VU}S$ is a distribution in the dual space D' (V) defined by

$$\langle \rho_{VU}S,\varphi\rangle = \langle S, E_{VU}\varphi\rangle$$

for all test functions $\varphi \in D(V)$.

Unless $U = V$, the restriction to V is neither injective nor surjective. Lack of surjectivity follows since distributions can blow up towards the boundary of V. For instance, if U = R and V = (0, 2), then the distribution

$$S(x) = \sum_{n=1}^{\infty} n\delta\left(x - \frac{1}{n}\right)$$

is in D' (V) but admits no extension to D' (U).

Support of a Distribution

Let $S \in D'$ (U) be a distribution on an open set U. Then S is said to vanish on an open set V of U if S lies in the kernel of the restriction map ρ_{VU}. Explicitly S vanishes on V if

$$\langle S,\varphi\rangle = 0$$

for all test functions $\varphi \in C^\infty(U)$ with support in V. Let V be a maximal open set on which the distribution S vanishes; i.e., V is the union of every open set on which S vanishes. The support of S is the complement of V in U. Thus

$$\operatorname{supp} S = U - \bigcup \{V \mid \rho_{VU} S = 0\}.$$

The distribution S has compact support if its support is a compact set. Explicitly, S has compact support if there is a compact subset K of U such that for every test function φ whose support is completely outside of K, we have $S(\varphi) = 0$. Compactly supported distributions define continuous linear functionals on the space $C^\infty(U)$; the topology on $C^\infty(U)$ is defined such that a sequence of test functions φ_k converges to 0 if and only if all derivatives of φ_k converge uniformly to 0 on every compact subset of U. Conversely, it can be shown that every continuous linear functional on this space defines a distribution of compact support. The embedding of $C_c^\infty(U)$ into $C^\infty(U)$, where the spaces are given their respective topologies, is continuous and has dense image. Thus compactly supported distributions can be identified with those distributions that can be extended from $C_c^\infty(U)$ to $C^\infty(U)$.

Tempered Distributions and Fourier Transform

By using a larger space of test functions, one can define the tempered distributions, a subspace of $D'(R^n)$. These distributions are useful if one studies the Fourier transform in generality: all tempered distributions have a Fourier transform, but not all distributions have one.

The space of test functions employed here, the so-called Schwartz space $S(R^n)$, is the function space of all infinitely differentiable functions that are rapidly decreasing at infinity along with all partial derivatives. Thus $\varphi : R^n \to R$ is in the Schwartz space provided that any derivative of φ, multiplied with any power of $|x|$, converges towards 0 for $|x| \to \infty$ These functions form a complete topological vector space with a suitably defined family of seminorms. More precisely, let

$$p_{\alpha,\beta}(\varphi) = \sup_{x \in R^n} |x^\alpha D^\beta \varphi(x)|$$

for α, β multi-indices of size n. Then φ is a Schwartz function if all the values

$$p_{\alpha,\beta}(\varphi) < \infty.$$

The family of seminorms $p_{\alpha,\beta}$ defines a locally convex topology on the Schwartz-space. The seminorms are, in fact, norms on the Schwartz space, since Schwartz functions are smooth. The Schwartz space is metrizable and complete. Because the Fourier transform changes differentiation by x^α into multiplication by x^α and vice-versa, this symmetry implies that the Fourier transformations of a Schwartz function is also a Schwartz function.

The space of tempered distributions is defined as the (continuous) dual of the Schwartz space. In other words, a distribution F is a tempered distribution if and only if

$$\lim_{m \to \infty} F(\varphi_m) = 0.$$

is true whenever,

$$\lim_{m \to \infty} p_{\alpha,\beta}(\varphi_m) = 0$$

holds for all multi-indices α, β.

The derivative of a tempered distribution is again a tempered distribution. Tempered distributions generalize the bounded (or slow-growing) locally integrable functions; all distributions with compact support and all square-integrable functions are tempered distributions. All locally integrable functions f with at most polynomial growth, i.e. such that $f(x) = O(|x|^r)$ for some r, are tempered distributions. This includes all functions in $L^p(\mathbb{R}^n)$ for $p \geq 1$.

The *tempered distributions* can also be characterized as *slowly growing*. This characterization is *dual* to the *rapidly falling* behaviour, e.g. $\propto |x|^n \cdot \exp(-x^2)$, of the test functions.

To study the Fourier transform, it is best to consider *complex*-valued test functions and complex-linear distributions. The ordinary continuous Fourier transform F yields then an automorphism of Schwartz function space, and we can define the Fourier transform of the tempered distribution S by $(FS)(\psi) = S(F\psi)$ for every test function ψ. FS is thus again a tempered distribution. The Fourier transform is a continuous, linear, bijective operator from the space of tempered distributions to itself. This operation is compatible with differentiation in the sense that

$$F\frac{dS}{dx} = ixFS$$

and also with convolution: if S is a tempered distribution and ψ is a *slowly increasing* infinitely differentiable function on R^n (meaning that all derivatives of ψ grow at most as fast as polynomials), then $S\psi$ is again a tempered distribution and

$$F(S\psi) = FS * F\psi$$

is the convolution of FS and $F\psi$. In particular, the Fourier transform of the unity function is the δ distribution.

Convolution

Under some circumstances, it is possible to define the convolution of a function with a distribution, or even the convolution of two distributions.

Convolution of a test function with a distribution: If $f \in D(R^n)$ is a compactly supported smooth test function, then convolution with f,

$$\begin{cases} C_f : \mathrm{D}(\mathrm{R}^n) \to \mathrm{D}(\mathrm{R}^n) \\ C_f g \mapsto f * g \end{cases}$$

defines a linear operator which is continuous with respect to the LF space topology on $D(R^n)$.

Convolution of f with a distribution $S \in D'(R^n)$ can be defined by taking the transpose of C_f relative to the duality pairing of $D(R^n)$ with the space $D'(R^n)$ of distributions (Trèves 1967, Chapter 27). If $f, g, \varphi \in D(R^n)$, then by Fubini's theorem

$$\left\langle C_f g, \varphi \right\rangle = \int_{\mathrm{R}^n} \varphi(x) \int_{\mathrm{R}^n} f(x-y) g(y) dy dx = \left\langle g, C_{\tilde{f}} \varphi \right\rangle$$

where $\tilde{f}(x) = f(-x)$. Extending by continuity, the convolution of f with a distribution S is defined by

$$\langle f * S, \varphi \rangle = \left\langle S, \tilde{f} * \varphi \right\rangle$$

for all test functions $\varphi \in \mathrm{D}(\mathrm{R}^n)$.

An alternative way to define the convolution of a function f and a distribution S is to use the translation operator τ_x defined on test functions by

$$\tau_x \varphi(y) = \varphi(y - x)$$

and extended by the transpose to distributions in the obvious way (Rudin 1991, §6.29). The convolution of the compactly supported function f and the distribution S is then the function defined for each $x \in \mathrm{R}^n$ by

$$(f * S)(x) = \left\langle S, \tau_x \tilde{f} \right\rangle.$$

It can be shown that the convolution of a compactly supported function and a distribution is a smooth function. If the distribution S has compact support as well, then $f * S$ is a compactly supported function, and the Titchmarsh convolution theorem (Hörmander 1983, Theorem 4.3.3) implies that

$$\mathrm{ch}(f * S) = \mathrm{ch}\, f + \mathrm{ch}\, S$$

where *ch* denotes the convex hull.

Distribution of compact support: It is also possible to define the convolution of two distributions S and T on R^n, provided one of them has compact support. Informally, in order to define $S*T$ where T has compact support, the idea is to extend the definition of the convolution " to a linear operation on distributions so that the associativity formula

$$S * (T * \varphi) = (S * T) * \varphi$$

continues to hold for all test-functions φ. Hörmander (1983, §IV.2) proves the uniqueness of such an extension.

It is also possible to provide a more explicit characterization of the convolution of distributions (Trèves 1967, Chapter 27). Suppose that it is T that has compact support. For any test function φ in $D(\mathrm{R}^n)$, consider the function

$$\psi(x) = \langle T, \tau_{-x} \varphi \rangle.$$

It can be readily shown that this defines a smooth function of x, which moreover has compact support. The convolution of S and T is defined by

$$\langle S * T, \varphi \rangle = \langle S, \psi \rangle.$$

This generalizes the classical notion of convolution of functions and is compatible with differentiation in the following sense:

$$\partial^\alpha (S * T) = (\partial^\alpha S) * T = S * (\partial^\alpha T).$$

This definition of convolution remains valid under less restrictive assumptions about S and T.

Distributions as Derivatives of Continuous Functions

The formal definition of distributions exhibits them as a subspace of a very large space, namely the topological dual of D(U) (or S(R^d) for tempered distributions). It is not immediately clear from the definition how exotic a distribution might be. To answer this question, it is instructive to see distributions built up from a smaller space, namely the space of continuous functions. Roughly, any distribution is locally a (multiple) derivative of a continuous function. A precise version of this result, given below, holds for distributions of compact support, tempered distributions, and general distributions. Generally speaking, no proper subset of the space of distributions contains all continuous functions and is closed under differentiation. This says that distributions are not particularly exotic objects; they are only as complicated as necessary.

Tempered Distributions

If $f \in S'(\mathrm{R}^n)$ is a tempered distribution, then there exists a constant $C > 0$, and positive integers M and N such that for all Schwartz functions $\varphi \in S(\mathrm{R}^n)$

$$\langle f,\varphi\rangle \leq C\sum\nolimits_{|\alpha|\leq N,|\beta|\leq M} \sup_{x\in\mathrm{R}^n}\left|x^{\alpha}D^{\beta}\varphi(x)\right| = C\sum\nolimits_{|\alpha|\leq N,|\beta|\leq M} p_{\alpha,\beta}(\varphi).$$

This estimate along with some techniques from functional analysis can be used to show that there is a continuous slowly increasing function F and a multiindex α such that

$$f = D^{\alpha}F.$$

Compactly Supported Distributions

Let U be an open set, and K a compact subset of U. If f is a distribution supported on K, then there is a continuous function F compactly supported in U (possibly on a larger set than K itself) such that

$$f = D^{\alpha}F$$

for some multi-index α. This follows from the previously quoted result on tempered distributions by means of a localization argument.

Distributions with Point Support

If f has support at a single point $\{P\}$, then f is in fact a finite linear combination of distributional derivatives of the δ function at P. That is, there exists an integer m and complex constants a_α for multi-indices $|\alpha| \leq m$ such that

$$f = \sum_{|\alpha| \leq m} a_\alpha D^\alpha (\tau_P \delta)$$

where τ_P is the translation operator.

General Distributions

A version of the above theorem holds locally in the following sense (Rudin 1991). Let S be a distribution on U, then one can find for every multi-index α a continuous function g_α such that

$$S = \sum_\alpha D^\alpha g_\alpha$$

and that any compact subset K of U intersects the supports of only finitely many g_α; therefore, to evaluate the value of S for a given smooth function f compactly supported in U, we only need finitely many g_α; hence the infinite sum above is well-defined as a distribution. If the distribution S is of finite order, then one can choose g_α in such a way that only finitely many of them are nonzero.

Using Holomorphic Functions as Test Functions

The success of the theory led to investigation of the idea of hyperfunction, in which spaces of holomorphic functions are used as test functions. A refined theory has been developed, in particular Mikio Sato's algebraic analysis, using sheaf theory and several complex variables. This extends the range of symbolic methods that can be made into rigorous mathematics, for example Feynman integrals.

Problem of Multiplication

A possible limitation of the theory of distributions (and hyperfunctions) is that it is a purely linear theory, in the sense that the product of two distributions cannot consistently be defined (in general), as has been proved by Laurent Schwartz in the 1950s. For example, if p.v. $1/x$ is the distribution obtained by the Cauchy principal value

$$\left(\mathrm{p.v.}\frac{1}{x}\right)[\phi] = \lim_{\epsilon \to 0^+} \int_{|x| \geq \epsilon} \frac{\phi(x)}{x} dx$$

for all $\varphi \in S(\mathrm{R})$, and δ is the Dirac delta distribution then

$$(\delta\times x)\times \text{p.v.}\frac{1}{x}=0$$

but

$$\delta\times\left(x\times \text{p.v.}\frac{1}{x}\right)=\delta$$

so the product of a distribution by a smooth function (which is always well defined) cannot be extended to an associative product on the space of distributions.

Thus, nonlinear problems cannot be posed in general and thus not solved within distribution theory alone. In the context of quantum field theory, however, solutions can be found. In more than two spacetime dimensions the problem is related to the regularization of divergences. Here Henri Epstein and Vladimir Glaser developed the mathematically rigorous (but extremely technical) *causal perturbation theory*. This does not solve the problem in other situations. Many other interesting theories are non linear, like for example Navier–Stokes equations of fluid dynamics.

In view of this, several not entirely satisfactory theories of algebras of generalized functions have been developed, among which Colombeau's (simplified) algebra is maybe the most popular in use today.

A simple solution of the multiplication problem is dictated by the path integral formulation of quantum mechanics. Since this is required to be equivalent to the Schrödinger theory of quantum mechanics which is invariant under coordinate transformations, this property must be shared by path integrals. This fixes all products of distributions as shown by Kleinert & Chervyakov (2001). The result is equivalent to what can be derived from dimensional regularization (Kleinert & Chervyakov 2000).

Dirac Delta Function

In mathematics, the Dirac delta function, or δ function, is (informally) a generalized function on the real number line that is zero everywhere except at zero, with an integral of one over the entire real line. The delta function is sometimes thought of as an infinitely high, infinitely thin spike at the origin, with total area one under the spike, and physically represents an idealized point mass or point charge. It was introduced by theoretical physicist Paul Dirac. Dirac explicitly spoke of infinitely great values of his integrand. In the context of signal processing it is often referred to as the unit impulse symbol (or function). Its discrete analog is the Kronecker delta function which is usually defined on a finite domain and takes values 0 and 1.

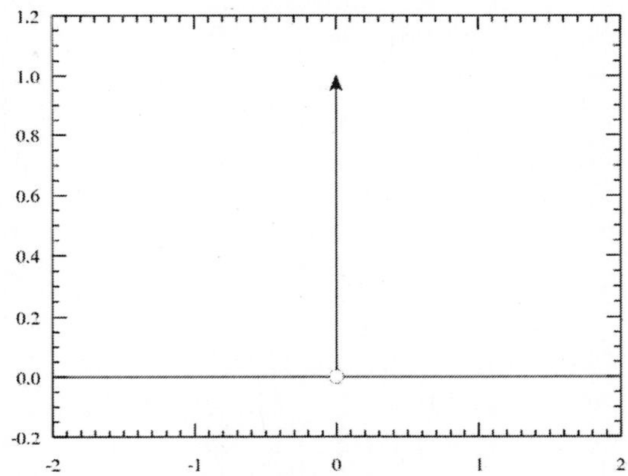

Figure: *Schematic representation of the Dirac delta function by a line surmounted by an arrow. The height of the arrow is usually used to specify the value of any multiplicative constant, which will give the area under the function. The other convention is to write the area next to the arrowhead.*

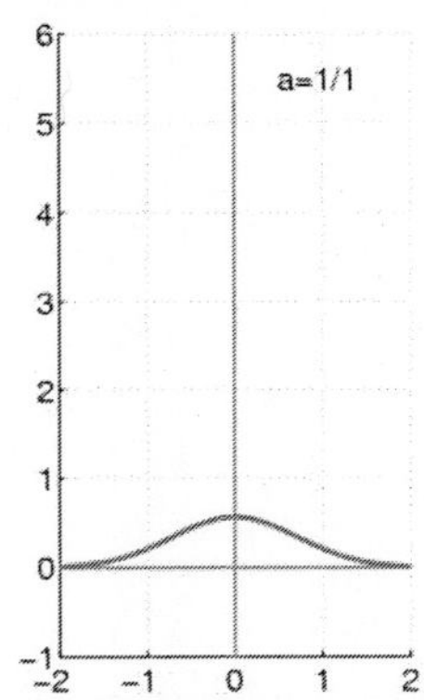

Figure: *The Dirac delta function as the limit (in the sense of distributions) of the sequence of zero-centred normal distributions*

$$\delta_a(x) = \frac{1}{a\sqrt{\pi}} e^{-x^2/a^2} \quad \textit{as} \ a \to 0 .$$

From a purely mathematical viewpoint, the Dirac delta is not strictly a function, because any extended-real function that is equal to zero everywhere but a single point must have total integral zero. The delta function only makes sense as a mathematical object when it appears inside an integral. While from this perspective the Dirac delta can usually be manipulated as though it were a function, formally it must be defined as a distribution that is also a measure. In many applications, the Dirac delta is regarded as a kind of limit (a weak limit) of a sequence of functions having a tall spike at the origin. The approximating functions of the sequence are thus "approximate" or "nascent" delta functions.

Overview

The graph of the delta function is usually thought of as following the whole x-axis and the positive y-axis. Despite its name, the delta function is not truly a function, at least not a usual one with range in real numbers. For example, the objects $f(x) = \delta(x)$ and $g(x) = 0$ are equal everywhere except at $x = 0$ yet have integrals that are different. According to Lebesgue integration theory, if f and g are functions such that $f = g$ almost everywhere, then f is integrable if and only if g is integrable and the integrals of f and g are identical. Rigorous treatment of the Dirac delta requires measure theory or the theory of distributions.

The Dirac delta is used to model a tall narrow spike function (an *impulse*), and other similar abstractions such as a point charge, point mass or electron point. For example, to calculate the dynamics of a baseball being hit by a bat, one can approximate the force of the bat hitting the baseball by a delta function. In doing so, one not only simplifies the equations, but one also is able to calculate the motion of the baseball by only considering the total impulse of the bat against the ball rather than requiring knowledge of the details of how the bat transferred energy to the ball.

In applied mathematics, the delta function is often manipulated as a kind of limit (a weak limit) of a sequence of functions, each member of which has a tall spike at the origin: for example, a sequence of Gaussian distributions centred at the origin with variance tending to zero.

History

Joseph Fourier presented what is now called the *Fourier integral theorem* in his treatise *Théorie analytique de la chaleur* in the form:

$$f(x) = \frac{1}{2\pi}\int_{-\infty}^{\infty} d\alpha f(\alpha) \int_{-\infty}^{\infty} dp \cos(px - p\alpha),$$

which is tantamount to the introduction of the δ-function in the form:

$$\delta(x - \alpha) = \frac{1}{2\pi}\int_{-\infty}^{\infty} dp \cos(px - p\alpha).$$

Later, Augustin Cauchy expressed the theorem using exponentials:

$$f(x) = \frac{1}{2\pi}\int_{-\infty}^{\infty} e^{ipx}\left(\int_{-\infty}^{\infty} e^{-ip\alpha} f(\alpha)\, d\alpha\right) dp.$$

Cauchy pointed out that in some circumstances the *order* of integration in this result was significant.

As justified using the theory of distributions, the Cauchy equation can be rearranged to resemble Fourier's original formulation and expose the δ-function as:

$$\begin{aligned} f(x) &= \frac{1}{2\pi}\int_{-\infty}^{\infty} e^{ipx}\left(\int_{-\infty}^{\infty} e^{-ip\alpha} f(\alpha)\, d\alpha\right) dp \\ &= \frac{1}{2\pi}\int_{-\infty}^{\infty}\left(\int_{-\infty}^{\infty} e^{ipx} e^{-ip\alpha}\, dp\right) f(\alpha)\, d\alpha = \int_{-\infty}^{\infty} \delta(x - \alpha) f(\alpha)\, d\alpha, \end{aligned}$$

where the δ-function is expressed as:

$$\delta(x - \alpha) = \frac{1}{2\pi}\int_{-\infty}^{\infty} e^{ip(x-\alpha)}\, dp.$$

A rigorous interpretation of the exponential form and the various limitations upon the function f necessary for its application extended over several centuries. The problems with a classical interpretation are explained as follows:

"The greatest drawback of the classical Fourier transformation is a rather narrow class of functions (originals) for which it can be effectively computed. Namely, it is necessary that these functions decrease sufficiently rapidly to zero (in the neighbourhood of infinity) in order to insure the existence of the Fourier integral. For example, the Fourier transform of such simple functions as polynomials does not exist in the classical sense. The extension of the classical Fourier transformation to distributions considerably enlarged the class of functions that could be transformed and this removed many obstacles."

Further developments included generalization of the Fourier integral, "beginning with Plancherel's pathbreaking L^2-theory (1910), continuing with Wiener's and Bochner's works (around 1930) and culminating with the amalgamation into L. Schwartz's theory of distributions (1945)...", and leading to the formal development of the Dirac delta function.

An infinitesimal formula for an infinitely tall, unit impulse delta function (infinitesimal version of Cauchy distribution) explicitly appears in an 1827 text of Augustin Louis Cauchy. Siméon Denis Poisson considered the issue in connection with the study of wave propagation as did Gustav Kirchhoff somewhat later. Kirchhoff and Hermann von Helmholtz also introduced the unit impulse as a limit of Gaussians, which also corresponded to Lord Kelvin's notion of a point heat source. At the end of the 19th century, Oliver Heaviside used formal Fourier series to manipulate the unit impulse. The Dirac delta function as such was introduced as a "convenient notation" by Paul Dirac in his influential 1930 book *Principles of Quantum Mechanics*. He called it the "delta function" since he used it as a continuous analogue of the discrete Kronecker delta.

Definitions

The Dirac delta can be loosely thought of as a function on the real line which is zero everywhere except at the origin, where it is infinite,

$$\delta(x) = \begin{cases} +\infty, & x = 0 \\ 0, & x \neq 0 \end{cases}$$

and which is also constrained to satisfy the identity

$$\int_{-\infty}^{\infty} \delta(x)dx = 1.$$

This is merely a heuristic characterization. The Dirac delta is not a function in the traditional sense as no function defined on the real numbers has these properties. The Dirac delta function can be rigorously defined either as a distribution or as a measure.

As a Measure

One way to rigorously define the delta function is as a measure, which accepts as an argument a subset A of the real line R, and returns $\delta(A) = 1$ if $0 \in A$, and $\delta(A) = 0$ otherwise. If the delta function is conceptualized as modelling an idealized point mass at 0, then $\delta(A)$ represents the mass contained in the set A. One may then define the integral against δ as the integral of a function against this mass distribution. Formally, the Lebesgue integral provides the necessary analytic device. The Lebesgue integral with respect to the measure δ satisfies

$$\int_{-\infty}^{\infty} f(x)\delta\{dx\} = f(0)$$

for all continuous compactly supported functions f. The measure δ is not absolutely continuous with respect to the Lebesgue measure — in fact, it is a singular measure. Consequently, the delta measure has no Radon–Nikodym derivative — no true function for which the property

$$\int_{-\infty}^{\infty} f(x)\delta(x)dx = f(0)$$

holds. As a result, the latter notation is a convenient abuse of notation, and not a standard (Riemann or Lebesgue) integral.

As a probability measure on R, the delta measure is characterized by its cumulative distribution function, which is the unit step function

$$H(x) = \begin{cases} 1 & \text{if } x \geq 0 \\ 0 & \text{if } x < 0. \end{cases}$$

This means that $H(x)$ is the integral of the cumulative indicator function $1_{(-\infty,\, x]}$ with respect to the measure δ; to wit,

$$H(x) = \int_{\mathrm{R}} 1_{(-\infty,x]}(t)\delta\{dt\} = \delta(-\infty,x].$$

Thus in particular the integral of the delta function against a continuous function can be properly understood as a Stieltjes integral:

$$\int_{-\infty}^{\infty} f(x)\delta\{dx\} = \int_{-\infty}^{\infty} f(x)dH(x).$$

All higher moments of δ are zero. In particular, characteristic function and moment generating function are both equal to one.

As a Distribution

In the theory of distributions a generalized function is thought of not as a function itself, but only in relation to how it affects other functions when it is "integrated" against them. In keeping with this philosophy, to define the delta function properly, it is enough to say what the "integral" of the delta function against a sufficiently "good" test function is. If the delta function is already understood as a measure, then the Lebesgue integral of a test function against that measure supplies the necessary integral.

A typical space of test functions consists of all smooth functions on R with compact support. As a distribution, the Dirac delta is a linear functional on the space of test functions and is defined by

$$\delta[\varphi] = \varphi(0)$$

for every test function φ.

For δ to be properly a distribution, it must be "continuous" in a suitable sense. In general, for a linear functional S on the space of test functions to define a distribution, it is necessary and sufficient that, for every positive integer N there is an integer M_N and a constant C_N such that for every test function φ, one has the inequality

$$|S[\phi]| \leq C_N \sum_{k=0}^{M_N} \sup_{x\in[-N,N]} |\phi^{(k)}(x)|.$$

With the δ distribution, one has such an inequality (with $C_N = 1$) with $M_N = 0$ for all N. Thus δ is a distribution of order zero. It is, furthermore, a distribution with compact support (the support being {0}).

The delta distribution can also be defined in a number of equivalent ways. For instance, it is the distributional derivative of the Heaviside step function. This means that, for every test function φ, one has

$$\delta[\phi] = -\int_{-\infty}^{\infty} \phi'(x) H(x)\, dx.$$

Intuitively, if integration by parts were permitted, then the latter integral should simplify to

$$\int_{-\infty}^{\infty} \phi(x)H'(x)dx = \int_{-\infty}^{\infty} \phi(x)\delta(x)dx,$$

and indeed, a form of integration by parts is permitted for the Stieltjes integral, and in that case one does have

$$-\int_{-\infty}^{\infty} \phi'(x)H(x)dx = \int_{-\infty}^{\infty} \phi(x)dH(x).$$

In the context of measure theory, the Dirac measure gives rise to a distribution by integration. Conversely, equation (1) defines a Daniell integral on the space of all compactly supported continuous functions φ which, by the Riesz representation theorem, can be represented as the Lebesgue integral of φ with respect to some Radon measure.

Generalizations

The delta function can be defined in n-dimensional Euclidean space R^n as the measure such that

$$\int_{\mathrm{R}^n} f(\mathrm{x})\delta\{d\mathrm{x}\} = f(0)$$

for every compactly supported continuous function f. As a measure, the n-dimensional delta function is the product measure of the 1-dimensional delta functions in each variable separately. Thus, formally, with $\mathrm{x} = (x_1, x_2, \ldots, x_n)$, one has

$$\delta(\mathrm{x}) = \delta(x_1)\delta(x_2)\ldots\delta(x_n).$$

The delta function can also be defined in the sense of distributions exactly as above in the one-dimensional case. However, despite widespread use in engineering contexts, (2) should be manipulated with care, since the product of distributions can only be defined under quite narrow circumstances.

The notion of a Dirac measure makes sense on any set whatsoever. Thus if X is a set, $x_0 \in X$ is a marked point, and Σ is any sigma algebra of subsets of X, then the measure defined on sets $A \in \Sigma$ by

$$\delta_{x_0}(A) = \begin{cases} 1 & \text{if } \mathrm{x}_0 \in \mathrm{A} \\ 0 & \text{if } \mathrm{x}_0 \notin \mathrm{A} \end{cases}$$

is the delta measure or unit mass concentrated at x_0.

Another common generalization of the delta function is to a differentiable manifold where most of its properties as a distribution can also be exploited because of the differentiable structure. The delta function on a manifold M centred at the point $x_0 \in M$ is defined as the following distribution:

$$\delta_{x_0}[\phi] = \phi(x_0)$$

for all compactly supported smooth real-valued functions φ on M. A common special case of this construction is when M is an open set in the Euclidean space R^n.

On a locally compact Hausdorff space X, the Dirac delta measure concentrated at a point x is the Radon measure associated with the Daniell integral (3) on compactly supported continuous functions φ. At this level of generality, calculus as such is no longer possible, however a variety of techniques from abstract analysis are available. For instance, the mapping $x_0 \mapsto \delta_{x_0}$ is a continuous embedding of X into the space of finite Radon measures on X, equipped with its vague topology. Moreover, the convex hull of the image of X under this embedding is dense in the space of probability measures on X.

Properties

Scaling and Symmetry: The delta function satisfies the following scaling property for a non-zero scalar α:

$$\int_{-\infty}^{\infty} \delta(\alpha x)dx = \int_{-\infty}^{\infty} \delta(u)\frac{du}{|\alpha|} = \frac{1}{|\alpha|}$$

and so

$$\delta(\alpha x) = \frac{\delta(x)}{|\alpha|}.$$

In particular, the delta function is an even distribution, in the sense that

$$\delta(-x) = \delta(x)$$

which is homogeneous of degree "1.

Algebraic Properties

The distributional product of δ with x is equal to zero:

$$x\delta(x) = 0.$$

Conversely, if $xf(x) = xg(x)$, where f and g are distributions, then

$$f(x) = g(x) + c\delta(x)$$

for some constant c.

Translation

The integral of the time-delayed Dirac delta is given by:

$$\int_{-\infty}^{\infty} f(t)\delta(t-T)dt = f(T).$$

This is sometimes referred to as the *sifting property* or the *sampling property*. The delta function is said to "sift out" the value at $t = T$.

It follows that the effect of convolving a function $f(t)$ with the time-delayed Dirac delta is to time-delay $f(t)$ by the same amount:

$$(f(t) * \delta(t-T)) \stackrel{\text{def}}{=} \int_{-\infty}^{\infty} f(\tau)\cdot\delta(t-T-\tau)d\tau$$

$$= \int_{-\infty}^{\infty} f(\tau)\cdot\delta(\tau-(t-T))d\tau$$

(using (4): $\delta(-x) = \delta(x)$)

$$= f(t-T).$$

This holds under the precise condition that f be a tempered distribution. As a special case, for instance, we have the identity (understood in the distribution sense)

$$\int_{-\infty}^{\infty} \delta(\xi - x)\delta(x-\eta)dx = \delta(\xi-\eta).$$

Composition with a Function

More generally, the delta distribution may be composed with a smooth function $g(x)$ in such a way that the familiar change of variables formula holds, that

$$\int_{\mathrm{R}} \delta\big(g(x)\big) f\big(g(x)\big) \mid g'(x) \mid dx = \int_{g(\mathrm{R})} \delta(u) f(u) du$$

provided that g is a continuously differentiable function with $g2$ nowhere zero. That is, there is a unique way to assign meaning to the distribution $\delta^\circ g$ so that this identity holds for all compactly supported test functions f. This distribution satisfies $\delta(g(x)) = 0$ if g is nowhere zero, and otherwise if g has a real root at x_0, then

$$\delta(g(x)) = \frac{\delta(x - x_0)}{|g'(x_0)|}.$$

It is natural therefore to *define* the composition $\delta(g(x))$ for continuously differentiable functions g by

$$\delta(g(x)) = \sum_i \frac{\delta(x - x_i)}{|g'(x_i)|}$$

where the sum extends over all roots of $g(x)$, which are assumed to be simple. Thus, for example

$$\delta\left(x^2 - \alpha^2\right) = \frac{1}{2|\alpha|}\left[\delta\left(x + \alpha\right) + \delta\left(x - \alpha\right)\right].$$

In the integral form the generalized scaling property may be written as

$$\int_{-\infty}^{\infty} f(x)\delta(g(x))dx = \sum_i \frac{f(x_i)}{|g'(x_i)|}.$$

Properties in N Dimensions

The delta distribution in an n-dimensional space satisfies the following scaling property instead:

$$\delta(\alpha \mathrm{x}) = |\alpha|^{-n}\ \delta(\mathrm{x})$$

so that δ is a homogeneous distribution of degree $-n$. Under any reflection or rotation ρ, the delta function is invariant:

$$\delta(\rho \mathrm{x}) = \delta(\mathrm{x}).$$

As in the one-variable case, it is possible to define the composition of δ with a bi-Lipschitz function g: $\mathrm{R}^n \to \mathrm{R}^n$ uniquely so that the identity

$$\int_{\mathrm{R}^n} \delta(g(\mathrm{x}))f(g(\mathrm{x}))|\det g'(\mathrm{x})|d\mathrm{x} = \int_{g(\mathrm{R}^n)} \delta(\mathrm{u})f(\mathrm{u})d\mathrm{u}$$

for all compactly supported functions f.

Using the coarea formula from geometric measure theory, one can also define the composition of the delta function with a submersion from one Euclidean space to another one of different dimension; the result is a type of current. In the special case of a continuously differentiable function g: $\mathrm{R}^n \to \mathrm{R}$ such that the gradient of g is nowhere zero, the following identity holds

$$\int_{\mathrm{R}^n} f(\mathrm{x})\delta(g(\mathrm{x}))d\mathrm{x} = \int_{g^{-1}(0)} \frac{f(\mathrm{x})}{|\nabla g|} d\sigma(\mathrm{x})$$

where the integral on the right is over $g^{-1}(0)$, the $n - 1$ dimensional surface defined by $g(\mathrm{x}) = 0$ with respect to the Minkowski content measure. This is known as a simple layer integral.

More generally, if S is a smooth hypersurface of R^n, then we can associated to S the distribution that integrates any compactly supported smooth function g over S:

$$\delta_S[g] = \int_S g(\mathrm{s})d\sigma(\mathrm{s})$$

where σ is the hypersurface measure associated to S. This generalization is associated with the potential theory of simple layer potentials on S. If D is a domain in R^n with smooth boundary S, then δ_S is equal to the normal derivative of the indicator function of D in the distribution sense:

$$-\int_{\mathrm{R}^n} g(\mathrm{x}) \frac{\partial 1_D(\mathrm{x})}{\partial n} d\mathrm{x} = \int_S g(\mathrm{s})\, d\sigma(\mathrm{s}),$$

where n is the outward normal.

Fourier Transform

The delta function is a tempered distribution, and therefore it has a well-defined Fourier transform. Formally, one finds

$$\hat{\delta}(\xi) = \int_{-\infty}^{\infty} e^{-2\pi i x \xi}\delta(x)dx = 1.$$

Properly speaking, the Fourier transform of a distribution is defined by imposing self-adjointness of the Fourier transform under the duality pairing $\langle \cdot, \cdot \rangle$ of tempered distributions with Schwartz functions. Thus $\hat{\delta}$ is defined as the unique tempered distribution satisfying

$$\langle \hat{\delta}, \phi \rangle = \langle \delta, \hat{\phi} \rangle$$

for all Schwartz functions φ. And indeed it follows from this that $\hat{\delta} = 1$.

As a result of this identity, the convolution of the delta function with any other tempered distribution S is simply S:

$$S * \delta = S.$$

That is to say that δ is an identity element for the convolution on tempered distributions, and in fact the space of compactly supported distributions under convolution is an associative algebra with identity the delta function. This property is fundamental in signal processing, as convolution with a tempered distribution is a linear time-invariant system, and applying the linear time-invariant system measures its impulse response. The impulse response can be computed to any desired degree of accuracy by choosing a suitable approximation for δ, and once it is known, it characterizes the system completely.

The inverse Fourier transform of the tempered distribution $f(\partial) = 1$ is the delta function. Formally, this is expressed

$$\int_{-\infty}^{\infty} 1 \cdot e^{2\pi i x \xi}\, d\xi = \delta(x)$$

and more rigorously, it follows since

$$\langle 1, f^{\vee} \rangle = f(0) = \langle \delta, f \rangle$$

for all Schwartz functions f.

In these terms, the delta function provides a suggestive statement of the orthogonality property of the Fourier kernel on R. Formally, one has

$$\int_{-\infty}^{\infty} e^{i2\pi\xi_1 t}\left[e^{i2\pi\xi_2 t}\right]^* dt = \int_{-\infty}^{\infty} e^{-i2\pi(\xi_2-\xi_1)t}\, dt = \delta(\xi_2 - \xi_1).$$

This is, of course, shorthand for the assertion that the Fourier transform of the tempered distribution

$$f(t) = e^{i2\pi\xi_1 t}$$

is

$$\hat{f}(\xi_2) = \delta(\xi_1 - \xi_2)$$

which again follows by imposing self-adjointness of the Fourier transform.

By analytic continuation of the Fourier transform, the Laplace transform of the delta function is found to be

$$\int_0^{\infty} \delta(t-a)e^{-st}\, dt = e^{-sa}.$$

Distributional Derivatives

The distributional derivative of the Dirac delta distribution is the distribution δ' defined on compactly supported smooth test functions φ by

$$\delta'[\varphi] = -\delta[\varphi'] = -\varphi'(0).$$

The first equality here is a kind of integration by parts, for if δ were a true function then

$$\int_{-\infty}^{\infty} \delta'(x)\varphi(x)dx = -\int_{-\infty}^{\infty} \delta(x)\varphi'(x)dx.$$

The k-th derivative of δ is defined similarly as the distribution given on test functions by

$$\delta^{(k)}[\varphi] = (-1)^k \varphi^{(k)}(0).$$

In particular δ is an infinitely differentiable distribution.

The first derivative of the delta function is the distributional limit of the difference quotients:

$$\delta'(x) = \lim_{h\to 0} \frac{\delta(x+h) - \delta(x)}{h}.$$

More properly, one has

$$\delta' = \lim_{h\to 0} \frac{1}{h}(\tau_h \delta - \delta)$$

where τ_h is the translation operator, defined on functions by $\tau_h\varphi(x) = \varphi(x+h)$, and on a distribution S by

$$(\tau_h S)[\varphi] = S[\tau_{-h}\varphi].$$

In the theory of electromagnetism, the first derivative of the delta function represents a point magnetic dipole situated at the origin. Accordingly, it is referred to as a dipole or the doublet function.

The derivative of the delta function satisfies a number of basic properties, including:

$$\frac{d}{dx}\delta(-x) = -\frac{d}{dx}\delta(x)$$

$$x\delta'(x) = -\delta(x).$$

Furthermore, the convolution of δ' with a compactly supported smooth function f is

$$\delta' * f = \delta * f' = f',$$

which follows from the properties of the distributional derivative of a convolution.

Higher Dimensions

More generally, on an open set U in the n-dimensional Euclidean space R^n, the Dirac delta distribution centred at a point $a \in U$ is defined by

$$\delta_a[\phi] = \phi(a)$$

for all $\varphi \in S(U)$, the space of all smooth compactly supported functions on U. If $\alpha = (\alpha_1, \ldots, \alpha_n)$ is any multi-index and ∂^α denotes the associated mixed partial derivative operator, then the αth derivative $\partial^\alpha\delta_a$ of δ_a is given by

$$\left\langle \partial^\alpha \delta_a, \varphi \right\rangle = (-1)^{|\alpha|} \left\langle \delta_a, \partial^\alpha \varphi \right\rangle = (-1)^{|\alpha|} \partial^\alpha \varphi(x)\Big|_{x=a} \text{ for all } \varphi \in S(U).$$

That is, the αth derivative of δ_a is the distribution whose value on any test function φ is the αth derivative of φ at a (with the appropriate positive or negative sign).

The first partial derivatives of the delta function are thought of as double layers along the coordinate planes. More generally, the normal derivative of a simple layer supported on a surface is a double layer supported on that surface, and represents a laminar magnetic monopole. Higher derivatives of the delta function are known in physics as multipoles.

Higher derivatives enter into mathematics naturally as the building blocks for the complete structure of distributions with point support. If S is any distribution on U supported on the set $\{a\}$ consisting of a single point, then there is an integer m and coefficients c_α such that

$$S = \sum_{|\alpha| \le m} c_\alpha \partial^\alpha \delta_a.$$

Representations of the Delta Function

The delta function can be viewed as the limit of a sequence of functions

$$\delta(x) = \lim_{\varepsilon \to 0^+} \eta_\varepsilon(x),$$

where $\eta_\varepsilon(x)$ is sometimes called a nascent delta function. This limit is meant in a weak sense: either that

$$\lim_{\varepsilon \to 0^+} \int_{-\infty}^{\infty} \eta_\varepsilon(x) f(x) dx = f(0)$$

for all continuous functions f having compact support, or that this limit holds for all smooth functions f with compact support. The difference between these two slightly different modes of weak convergence is often subtle: the former is convergence in the vague topology of measures, and the latter is convergence in the sense of distributions.

Approximations to the Identity

Typically a nascent delta function η_ε can be constructed in the following manner. Let η be an absolutely integrable function on R of total integral 1, and define

$$\eta_\varepsilon(x) = \varepsilon^{-1} \eta\left(\frac{x}{\varepsilon}\right).$$

In n dimensions, one uses instead the scaling

$$\eta_\varepsilon(x) = \varepsilon^{-n} \eta\left(\frac{x}{\varepsilon}\right).$$

Then a simple change of variables shows that η_ε also has integral 1. One shows easily that (5) holds for all continuous compactly supported functions f, and so η_ε converges weakly to δ in the sense of measures.

The η_ε constructed in this way are known as an approximation to the identity. This terminology is because the space $L^1(\mathrm{R})$ of absolutely integrable functions is closed under the operation of convolution of functions: $f*g \in L^1(\mathrm{R})$ whenever f and g are in $L^1(\mathrm{R})$. However, there is no identity in $L^1(\mathrm{R})$ for the convolution product: no element h such that $f*h = f$ for all f. Nevertheless, the sequence η_ε does approximate such an identity in the sense that

$$f * \eta_\varepsilon \to f \quad \text{as } \varepsilon \to 0.$$

This limit holds in the sense of mean convergence (convergence in L^1). Further conditions on the η_ε, for instance that it be a mollifier associated to a compactly supported function, are needed to ensure pointwise convergence almost everywhere.

If the initial $\eta = \eta_1$ is itself smooth and compactly supported then the sequence is called a mollifier. The standard mollifier is obtained by choosing η to be a suitably normalized bump function, for instance

$$\eta(x) = \begin{cases} e^{-\frac{1}{1-|x|^2}} & \text{if } |x| < 1 \\ 0 & \text{if } |x| \geq 1. \end{cases}$$

In some situations such as numerical analysis, a piecewise linear approximation to the identity is desirable. This can be obtained by taking η_1 to be a hat function. With this choice of η_1, one has

$$\eta_\varepsilon(x) = \varepsilon^{-1} \max\left(1 - \left|\frac{x}{\varepsilon}\right|, 0\right)$$

which are all continuous and compactly supported, although not smooth and so not a mollifier.

Probabilistic Considerations

In the context of probability theory, it is natural to impose the additional condition that the initial η_1 in an approximation to the identity should be positive, as such a function then represents a probability distribution. Convolution with a probability distribution is sometimes favourable because it does not result in overshoot or undershoot, as the output is a convex combination of the input values, and thus falls between the maximum and minimum of the input function. Taking η_1 to be any probability distribution at all, and letting $\eta_\varepsilon(x) = \eta_1(x/\varepsilon)/\varepsilon$ as above will give rise to an approximation to the identity. In general this converges more rapidly to a delta function if, in addition, η has mean 0 and has small higher moments. For instance, if η_1 is the uniform distribution on $[-1/2, 1/2]$, also known as the rectangular function, then:

$$\eta_\varepsilon(x) = \frac{1}{\varepsilon}\operatorname{rect}\left(\frac{x}{\varepsilon}\right) = \begin{cases} \frac{1}{\varepsilon}, & -\frac{\varepsilon}{2} < x < \frac{\varepsilon}{2} \\ 0, & \text{otherwise.} \end{cases}$$

Another example is with the Wigner semicircle distribution

$$\eta_\varepsilon(x) = \begin{cases} \frac{2}{\pi\varepsilon^2}\sqrt{\varepsilon^2 - x^2}, & -\varepsilon < x < \varepsilon \\ 0, & \text{otherwise} \end{cases}$$

This is continuous and compactly supported, but not a mollifier because it is not smooth.

Semigroups

Nascent delta functions often arise as convolution semigroups. This amounts to the further constraint that the convolution of η_ε with η_δ must satisfy

$$\eta_\varepsilon * \eta_\delta = \eta_{\varepsilon+\delta}$$

for all ε, $\delta > 0$. Convolution semigroups in L^1 that form a nascent delta function are always an approximation to the identity in the above sense, however the semigroup condition is quite a strong restriction.

In practice, semigroups approximating the delta function arise as fundamental solutions or Green's functions to physically motivated elliptic or parabolic partial differential equations. In the context of applied mathematics, semigroups arise as the output of a linear time-invariant system. Abstractly, if A is a linear operator acting on functions of x, then a convolution semigroup arises by solving the initial value problem

$$\begin{cases} \dfrac{\partial}{\partial t}\eta(t,x) = A\eta(t,x), \quad t > 0 \\ \lim\limits_{t\to 0^+} \eta(t,x) = \delta(x) \end{cases}$$

in which the limit is as usual understood in the weak sense. Setting $\eta_\varepsilon(x) = \eta(\varepsilon, x)$ gives the associated nascent delta function.

Some examples of physically important convolution semigroups arising from such a fundamental solution include the following.

The heat kernel: The heat kernel, defined by

$$\eta_\varepsilon(x) = \frac{1}{\sqrt{2\pi\varepsilon}} e^{-\frac{x^2}{2\varepsilon}}$$

represents the temperature in an infinite wire at time $t > 0$, if a unit of heat energy is stored at the origin of the wire at time $t = 0$. This semigroup evolves according to the one-dimensional heat equation:

$$\frac{\partial u}{\partial t} = \frac{1}{2}\frac{\partial^2 u}{\partial x^2}.$$

In probability theory, $\eta_\varepsilon(x)$ is a normal distribution of variance ε and mean 0. It represents the probability density at time $t = \varepsilon$ of the position of a particle starting at the origin following a standard Brownian motion. In this context, the semigroup condition is then an expression of the Markov property of Brownian motion.

In higher dimensional Euclidean space R^n, the heat kernel is

$$\eta_\varepsilon = \frac{1}{(2\pi\varepsilon)^{n/2}} e^{-\frac{x \cdot x}{2\varepsilon}},$$

and has the same physical interpretation, *mutatis mutandis.* It also represents a nascent delta function in the sense that $\eta_\varepsilon \to \delta$ in the distribution sense as $\varepsilon \to 0$.

The Poisson kernel:

$$\eta_\varepsilon(x) = \frac{1}{\pi} \frac{\varepsilon}{\varepsilon^2 + x^2} = \int_{-\infty}^{\infty} e^{2\pi i \xi x - |\varepsilon\xi|} \, d\xi$$

is the fundamental solution of the Laplace equation in the upper half-plane. It represents the electrostatic potential in a semi-infinite plate whose potential along the edge is held at fixed at the delta function. The Poisson kernel is also closely related to the Cauchy distribution. This semigroup evolves according to the equation

$$\frac{\partial u}{\partial t} = -\left(-\frac{\partial^2}{\partial x^2}\right)^{\frac{1}{2}} u(t,x)$$

where the operator is rigorously defined as the Fourier multiplier

$$\mathcal{F}\left[\left(-\frac{\partial^2}{\partial x^2}\right)^{\frac{1}{2}} f\right](\xi) = |2\pi\xi| \mathcal{F}f(\xi).$$

Oscillatory Integrals

In areas of physics such as wave propagation and wave mechanics, the equations involved are hyperbolic and so may have more singular solutions. As a result, the nascent delta functions that arise as fundamental solutions of the associated Cauchy problems are generally oscillatory integrals. An example, which comes from a solution of the Euler–Tricomi equation of transonic gas dynamics, is the rescaled Airy function

$$\varepsilon^{-\frac{1}{3}}\operatorname{Ai}\left(x\varepsilon^{-\frac{1}{3}}\right).$$

Although using the Fourier transform, it is easy to see that this generates a semigroup in some sense, it is not absolutely integrable and so cannot define a semigroup in the above strong sense. Many nascent delta functions constructed as oscillatory integrals only converge in the sense of distributions (an example is the Dirichlet kernel below), rather than in the sense of measures.

Another example is the Cauchy problem for the wave equation in R^{1+1}:

$$c^{-2}\frac{\partial^2 u}{\partial t^2} - \Delta u = 0$$

$$u = 0, \quad \frac{\partial u}{\partial t} = \delta \qquad \text{for } t = 0.$$

The solution u represents the displacement from equilibrium of an infinite elastic string, with an initial disturbance at the origin.

Other approximations to the identity of this kind include the sinc function

$$\eta_\varepsilon(x) = \frac{1}{\pi x}\sin\left(\frac{x}{\varepsilon}\right) = \frac{1}{2\pi}\int_{-\frac{1}{\varepsilon}}^{\frac{1}{\varepsilon}} \cos(kx)\, dk$$

and the Bessel function

$$\eta_\varepsilon(x) = \frac{1}{\varepsilon}J_{\frac{1}{\varepsilon}}\left(\frac{x+1}{\varepsilon}\right).$$

Plane Wave Decomposition

One approach to the study of a linear partial differential equation

$$L[u] = f,$$

where L is a differential operator on R^n, is to seek first a fundamental solution, which is a solution of the equation

$$L[u] = \delta.$$

When L is particularly simple, this problem can often be resolved using the Fourier transform directly (as in the case of the Poisson kernel and heat kernel already mentioned). For more complicated operators, it is sometimes easier first to consider an equation of the form

$$L[u] = h$$

where h is a plane wave function, meaning that it has the form

$$h = h(x \cdot \xi)$$

for some vector ∂. Such an equation can be resolved (if the coefficients of L are analytic functions) by the Cauchy–Kovalevskaya theorem or (if the coefficients of L are constant) by quadrature. So, if the delta function can be decomposed into plane waves, then one can in principle solve linear partial differential equations.

Such a decomposition of the delta function into plane waves was part of a general technique first introduced essentially by Johann Radon, and then developed in this form by Fritz John (1955). Choose k so that $n + k$ is an even integer, and for a real number s, put

$$g(s) = \operatorname{Re}\left[\frac{-s^k \log(-is)}{k!(2\pi i)^n}\right] = \begin{cases} \dfrac{|s|^k}{4k!(2\pi i)^{n-1}} & n \text{ odd} \\[2ex] -\dfrac{|s|^k \log|s|}{k!(2\pi i)^n} & n \text{ even.} \end{cases}$$

Then δ is obtained by applying a power of the Laplacian to the integral with respect to the unit sphere measure $d\omega$ of $g(x \,.\, \partial)$ for ∂ in the unit sphere S^{n-1}:

$$\delta(x) = \Delta_x^{\frac{n+k}{2}} \int_{S^{n-1}} g(x \cdot \xi)\, d\omega_\xi .$$

The Laplacian here is interpreted as a weak derivative, so that this equation is taken to mean that, for any test function φ,

$$\varphi(x) = \int_{\mathbb{R}^n} \varphi(y)\, dy\, \Delta_x^{\frac{n+k}{2}} \int_{S^{n-1}} g((x - y) \cdot \xi)\, d\omega_\xi .$$

The result follows from the formula for the Newtonian potential (the fundamental solution of Poisson's equation). This is essentially a form of the inversion formula for the Radon transform, because it recovers the value of $\varphi(x)$ from its integrals over hyperplanes. For instance, if n is odd and $k = 1$, then the integral on the right hand side is

$$c_n \Delta_x^{\frac{n+1}{2}} \iint_{S^{n-1}} \varphi(y)\,|(y-x)\cdot\xi\,|d\omega_\xi\, dy = c_n \Delta_x^{\frac{n+1}{2}} \int_{S^{n-1}} d\omega_\xi \int_{-\infty}^{\infty} |p\,|\,R\varphi(\xi, p + x\cdot\xi)dp$$

where $R\varphi(\partial, p)$ is the Radon transform of φ:

$$R\varphi(\xi, p) = \int_{x\cdot\xi=p} f(x)d^{n-1}x.$$

An alternative equivalent expression of the plane wave decomposition, from Gel'fand & Shilov (1966–1968, I, §3.10), is

$$\delta(x) = \frac{(n-1)!}{(2\pi i)^n} \int_{S^{n-1}} (x\cdot\xi)^{-n}\, d\omega_\xi$$

for n even, and

$$\delta(x) = \frac{1}{2(2\pi i)^{n-1}} \int_{S^{n-1}} \delta^{(n-1)}(x\cdot\xi) d\omega_\xi$$

for n odd.

Fourier Kernels

In the study of Fourier series, a major question consists of determining whether and in what sense the Fourier series associated with a periodic function converges to the function. The nth partial sum of the Fourier series of a function f of period 2π is defined by convolution (on the interval $[-\pi,\pi]$) with the Dirichlet kernel:

$$D_N(x) = \sum_{n=-N}^{N} e^{inx} = \frac{\sin\left((N+\frac{1}{2})x\right)}{\sin(x/2)}.$$

Thus,

$$s_N(f)(x) = D_N * f(x) = \sum_{n=-N}^{N} a_n e^{inx}$$

where

$$a_n = \frac{1}{2\pi} \int_{-\pi}^{\pi} f(y) e^{-iny}\, dy.$$

A fundamental result of elementary Fourier series states that the Dirichlet kernel tends to the a multiple of the delta function as $N \to \infty$ This is interpreted in the distribution sense, that

$$s_N(f)(0) = \int_{\mathbb{R}} D_N(x) f(x)dx \to 2\pi f(0)$$

for every compactly supported *smooth* function f. Thus, formally one has

$$\delta(x) = \frac{1}{2\pi} \sum_{n=-\infty}^{\infty} e^{inx}$$

on the interval $[-\pi,\pi]$.

In spite of this, the result does not hold for all compactly supported *continuous* functions: that is D_N does not converge weakly in the sense of measures. The lack of convergence of the Fourier series has led to the introduction of a variety of summability methods in order to produce convergence. The method of Cesàro summation leads to the Fejér kernel

$$F_N(x) = \sum_{n=0}^{N} D_n(x) = \frac{1}{N}\left(\frac{\sin\frac{Nx}{2}}{\sin\frac{x}{2}}\right)^2.$$

The Fejér kernels tend to the delta function in a stronger sense that

$$\int_{\mathbb{R}} F_N(x) f(x)dx \to 2\pi f(0)$$

for every compactly supported *continuous* function f. The implication is that the Fourier series of any continuous function is Cesàro summable to the value of the function at every point.

Hilbert Space Theory

The Dirac delta distribution is a densely defined unbounded linear functional on the Hilbert space L^2 of square integrable functions. Indeed, smooth compactly support functions are dense in L^2, and the action of the delta distribution on such functions is well-defined. In many applications, it is possible to identify subspaces of L^2 and to give a stronger topology on which the delta function defines a bounded linear functional.

Sobolev Spaces

The Sobolev embedding theorem for Sobolev spaces on the real line R implies that any square-integrable function *f* such that

$$\| f \|_{H^1}^2 = \int_{-\infty}^{\infty} |\hat{f}(\xi)|^2 \, (1 + |\xi|^2) d\xi < \infty$$

is automatically continuous, and satisfies in particular

$$\delta[f] = | f(0) | < C \| f \|_{H^1} .$$

Thus δ is a bounded linear functional on the Sobolev space H^1. Equivalently δ is an element of the continuous dual space H^{-1} of H^1. More generally, in *n* dimensions, one has $\delta \in H^{-s}(\mathbb{R}^n)$ provided $s > n / 2$.

Spaces of Holomorphic Functions

In complex analysis, the delta function enters via Cauchy's integral formula which asserts that if *D* is a domain in the complex plane with smooth boundary, then

$$f(z) = \frac{1}{2\pi i} \oint_{\partial D} \frac{f(\zeta) d\zeta}{\zeta - z}, \quad z \in D$$

for all holomorphic functions *f* in *D* that are continuous on the closure of *D*. As a result, the delta function δ_z is represented on this class of holomorphic functions by the Cauchy integral:

$$\delta_z[f] = f(z) = \frac{1}{2\pi i} \oint_{\partial D} \frac{f(\zeta) d\zeta}{\zeta - z}.$$

More generally, let $H^2(\partial D)$ be the Hardy space consisting of the closure in $L^2(\partial D)$ of all holomorphic functions in *D* continuous up to the boundary of *D*. Then functions in $H^2(\partial D)$ uniquely extend to holomorphic functions in *D*, and the Cauchy integral formula continues to hold. In particular for $z \in D$, the delta function δ_z is a continuous linear functional on $H^2(\partial D)$. This is a special case of the situation in several complex variables in which, for smooth domains *D*, the Szegõ kernel plays the role of the Cauchy integral.

Resolutions of the Identity

Given a complete orthonormal basis set of functions $\{\varphi_n\}$ in a separable Hilbert space, for example, the normalized eigenvectors of a compact self-adjoint operator, any vector *f* can be expressed as:

$$f = \sum_{n=1}^{\infty} \alpha_n \varphi_n.$$

The coefficients $\{\alpha_n\}$ are found as:

$$\alpha_n = \langle \varphi_n, f \rangle,$$

which may be represented by the notation:

$$\alpha_n = \varphi_n^\dagger f,$$

a form of the bra-ket notation of Dirac. Adopting this notation, the expansion of f takes the dyadic form:

$$f = \sum_{n=1}^{\infty} \varphi_n \left(\varphi_n^\dagger f \right).$$

Letting I denote the identity operator on the Hilbert space, the expression

$$I = \sum_{n=1}^{\infty} \varphi_n \varphi_n^\dagger,$$

is called a resolution of the identity. When the Hilbert space is the space $L^2(D)$ of square-integrable functions on a domain D, the quantity:

$$\varphi_n \varphi_n^\dagger,$$

is an integral operator, and the expression for f can be rewritten as:

$$f(x) = \sum_{n=1}^{\infty} \int_D \left(\varphi_n(x) \varphi_n^*(\xi) \right) f(\xi) d\xi.$$

The right-hand side converges to f in the L^2 sense. It need not hold in a pointwise sense, even when f is a continuous function. Nevertheless, it is common to abuse notation and write

$$f(x) = \int \delta(x - \xi) f(\xi) d\xi,$$

resulting in the representation of the delta function:

$$\delta(x - \xi) = \sum_{n=1}^{\infty} \varphi_n(x) \varphi_n^*(\xi).$$

With a suitable rigged Hilbert space $(\Phi, L^2(D), \Phi^*)$ where $\Phi \subset L^2(D)$ contains all compactly supported smooth functions, this summation may converge in Φ^*, depending on the properties of the basis φ_n. In most cases of practical interest, the orthonormal basis comes from an integral or differential operator, in which case the series converges in the distribution sense.

Infinitesimal Delta Functions

Cauchy used an infinitesimal α to write down a unit impulse, infinitely tall and narrow Dirac-type delta function δ_α satisfying $\int F(x)\delta_\alpha(x) = F(0)$ in a number of articles in 1827. Cauchy defined an infinitesimal in Cours d'Analyze (1827) in terms of a sequence tending to zero. Namely, such a null sequence becomes an infinitesimal in Cauchy's and Lazare Carnot's terminology.

Modern set-theoretic approaches allow one to define infinitesimals via the ultrapower construction, where a null sequence becomes an infinitesimal in the sense of an equivalence class modulo a relation defined in terms of a suitable ultrafilter. The article by Yamashita (2007) contains a bibliography on modern Dirac delta functions in the context of an infinitesimal-enriched continuum provided by the hyperreals. Here the Dirac delta can be given by an actual function, having the property that for every real function F one has $\int F(x)\delta_\alpha(x) = F(0)$ as anticipated by Fourier and Cauchy.

Chapter 3

Nyquist–Shannon Sampling Theorem

The Nyquist–Shannon sampling theorem, after Harry Nyquist and Claude Shannon, in the literature more commonly referred to as the Nyquist sampling theorem or simply as the sampling theorem, is a fundamental result in the field of information theory, in particular telecommunications and signal processing. Sampling is the process of converting a signal (for example, a function of continuous time or space) into a numeric sequence (a function of discrete time or space). Shannon's version of the theorem states:

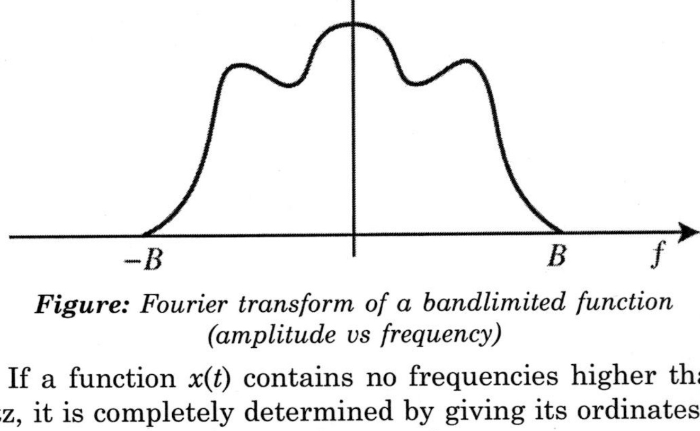

Figure: *Fourier transform of a bandlimited function (amplitude vs frequency)*

If a function $x(t)$ contains no frequencies higher than B hertz, it is completely determined by giving its ordinates at a series of points spaced $1/(2B)$ seconds apart.

In other words, a bandlimited function can be perfectly reconstructed from an infinite sequence of samples if the bandlimit, B, is no greater than ½ the sampling rate (samples per second). The theorem also leads to a formula for reconstruction of the original function from its samples. When the bandlimit is too high (or there is no bandlimit), the reconstruction exhibits imperfections known as aliasing. The Poisson summation formula provides a graphic understanding of aliasing and an alternative derivation of the theorem, using the perspective of the function's Fourier transform.

In practice of course, infinite sequences, perfect sampling, and perfect interpolation are all replaced by approximations that deviate from the mathematical ideal of perfect reconstruction.

Modern statements of the theorem are sometimes careful to explicitly state that $x(t)$ must contain no sinusoidal component at exactly frequency B, or that B must be strictly less than ½ the sample rate. And the theorem does not preclude the possibility of perfect reconstruction under special circumstances that do not satisfy the sample-rate criterion. It is a sufficient, but not necessary, condition.

To formalize the statements above, let $X(f)$ be the Fourier transform of bandlimited function $x(t)$:

$$X(f) \stackrel{\text{def}}{=} \int_{-\infty}^{\infty} x(t)\, e^{-i2\pi ft}\, \mathrm{d}t, \quad \text{and } X(f) = 0 \quad \text{for all } |f| > B.$$

When $x(t)$ is sampled uniformly at intervals of T seconds, the resultant sequence is denoted by x(nT), for all integer values of n. And the sample-rate (samples per second) is:

$$f_s \stackrel{\text{def}}{=} 1/T.$$

A sufficient condition to reconstruct $x(t)$ from its samples is $f_s > 2B$, and equivalently $B < f_s/2$. The two thresholds, $2B$ and $f_s/2$, are respectively called the Nyquist rate and Nyquist frequency. And respectively, they are attributes of $x(t)$ and of the sampling equipment. The condition described by these inequalities is called the Nyquist criterion, or sometimes the *Raabe condition*. The theorem is also applicable to functions of other domains, such as *space* in the case of a digitized image. All that changes is the units of measure for t, T, f, f_s, and B.

One way to ensure the Nyquist criterion is met (as an approximation) is by understanding the physics of the origination of $x(t)$ and selecting an appropriate rate. For example, sounds that are made by human voice normally contain relatively insignificant frequency components at or above 10 kHz. Sampling such an audio signal at 20k sam/sec or more provides an excellent approximation to meeting the criterion.

But often the sample-rate is pre-determined by other considerations, such as an industry standard format (e.g. 8k sam/sec). In that case, human voice should be filtered to remove frequency components above 4 kHz before being sampled. The type of filter required is a lowpass filter, and in this application it is called an anti-aliasing filter.

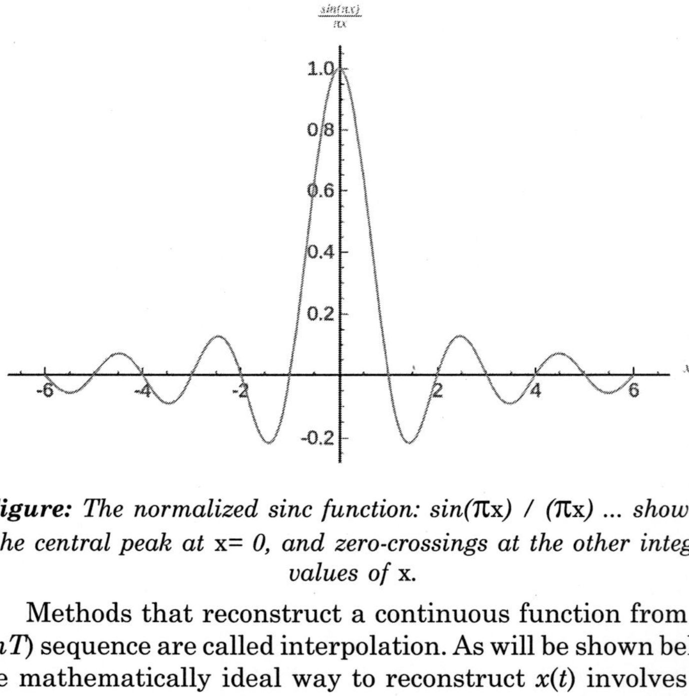

***Figure:** The normalized sinc function: sin(πx) / (πx) ... showing the central peak at x= 0, and zero-crossings at the other integer values of x.*

Methods that reconstruct a continuous function from the $x(nT)$ sequence are called interpolation. As will be shown below, the mathematically ideal way to reconstruct $x(t)$ involves the use of sinc functions, like that shown in figure. Each sample in the sequence is replaced by a sinc function centred on the time axis at the original location of the sample (nT), and the amplitude of the sinc function is scaled to the sample value, $x(nT)$. Then all the sinc functions are summed into a continuous function. A mathematically equivalent method is to convolve one sinc function with a series of Dirac delta pulses, weighted by the sample values. Neither method is numerically practical. Instead, some type of approximation of the sinc functions, finite in length, has to be used.

The imperfections attributable to the approximation are known as *interpolation error*.

Practical digital-to-analog converters produce neither scaled and delayed sinc functions nor ideal Dirac pulses. Instead they produce a piecewise-constant sequence of scaled and delayed rectangular pulses, usually followed by a "shaping filter" to clean up spurious high-frequency content.

Aliasing

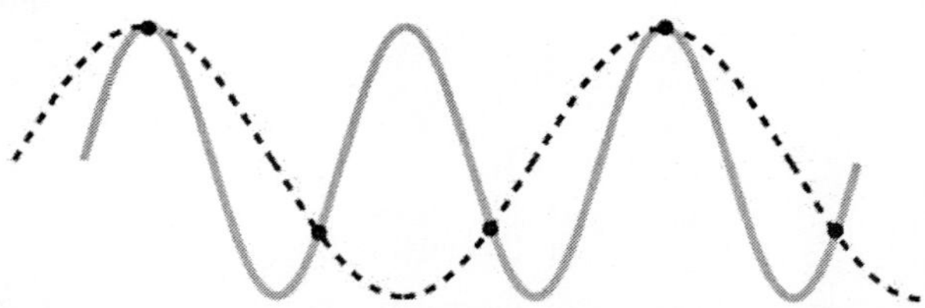

Figure: *The samples of several different sine waves can be identical, when at least one of them is at a frequency above half the sample rate.*

The Poisson summation formula shows that the samples, $x(nT)$, of function $x(t)$ are sufficient to create a periodic summation of function $X(f)$. The result is ($T=1/f_s$):

$$X_s(f) \stackrel{\text{def}}{=} \sum_{k=-\infty}^{\infty} X(f - kf_s) = \sum_{n=-\infty}^{\infty} \underbrace{T\cdot x(nT)}_{x[n]} \, e^{-i2\pi nTf},$$

which is a periodic function and its equivalent representation as a Fourier series, whose coefficients are $x[n]$. This function is also known as the discrete-time Fourier transform (DTFT). As depicted in Figures, copies of $X(f)$ are shifted by multiples of f_s and combined by addition.

If the Nyquist criterion is not satisfied, adjacent copies overlap, and it is not possible in general to discern an unambiguous $X(f)$. Any frequency component above $f_s/2$ is indistinguishable from a lower-frequency component, called an *alias*, associated with one of the copies. In such cases, the reconstruction technique described below produces the alias, rather than the original component.

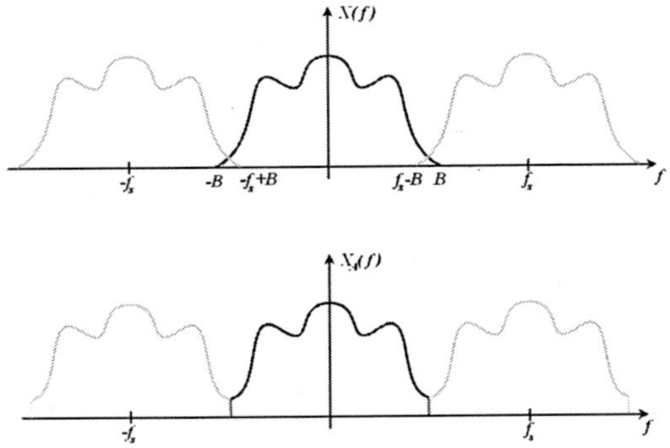

Figure: *X(f) (top blue) and $X_A(f)$ (bottom blue) are continuous Fourier transforms of two different functions, x(t) and $x_A(t)$ (not shown). When the functions are sampled at rate f_s, the images (green) are added to the original transforms (blue) when one examines the discrete-time Fourier transforms (DTFT) of the sequences. In this hypothetical example, the DTFTs are identical, which means the sampled sequences are identical, even though the original continuous pre-sampled functions are not. If these were audio signals, x(t) and $x_A(t)$ might not sound the same. But their samples (taken at rate f_s) are identical and would lead to identical reproduced sounds; thus $x_A(t)$ is an alias of x(t) at this sample rate. In this example (of a bandlimited function), such aliasing can be prevented by increasing f_s such that the green images in the top figure do not overlap the blue portion.*

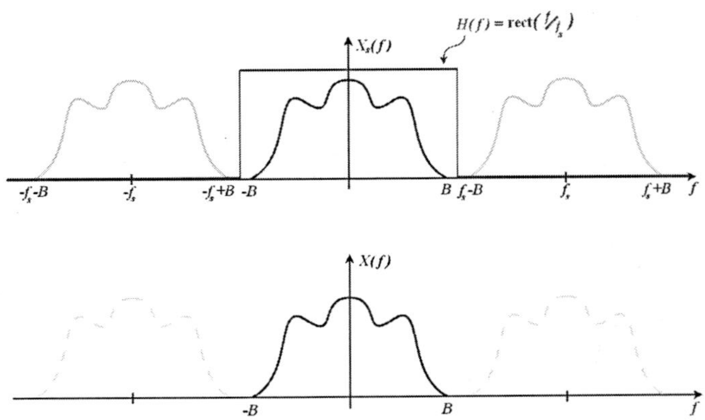

Figure: *Spectrum, $X_s(f)$, of a properly sampled bandlimited signal (blue) and the adjacent DTFT images (green) that do not overlap. A brick-wall low-pass filter, H(f), removes the images, leaves the original spectrum, X(f), and recovers the original signal from its samples.*

Derivation as a Special case of Poisson Summation

From previous figure, it is apparent that when there is no overlap of the copies (aka "images") of $X(f)$, the $k = 0$ term of $X_s(f)$ can be recovered by the product:

$$X(f) = H(f) \cdot X_s(f), \qquad \text{where:}$$

$$H(f) \stackrel{\text{def}}{=} \begin{cases} 1 & |f| < B \\ 0 & |f| > f_s - B. \end{cases}$$

At this point, the sampling theorem is proved, since $X(f)$ uniquely determines $x(t)$.

All that remains is to derive the formula for reconstruction. $H(f)$ need not be precisely defined in the region $[B, f_s - B]$ because $X_s(f)$ is zero in that region. However, the worst case is when $B = f_s/2$, the Nyquist frequency. A function that is sufficient for that and all less severe cases is:

$$H(f) = \text{rect}\left(\frac{f}{f_s}\right) = \begin{cases} 1 & |f| < \dfrac{f_s}{2} \\ 0 & |f| > \dfrac{f_s}{2}, \end{cases}$$

where rect(•) is the rectangular function. Therefore:

$$X(f) = \text{rect}\left(\frac{f}{f_s}\right) \cdot X_s(f)$$

$$= \text{rect}(Tf) \cdot \sum_{n=-\infty}^{\infty} T \cdot x(nT)\, e^{-i2\pi nTf}$$

(from Eq.1, above).

$$= \sum_{n=-\infty}^{\infty} x(nT) \cdot \underbrace{T \cdot \text{rect}(Tf) \cdot e^{-i2\pi nTf}}_{\mathcal{F}\left\{\text{sinc}\left(\frac{t-nT}{T}\right)\right\}}.$$

The inverse transform of both sides produces the Whittaker–Shannon interpolation formula:

$$x(t) = \sum_{n=-\infty}^{\infty} x(nT) \cdot \text{sinc}\left(\frac{t - nT}{T}\right),$$

which shows how the samples, $x(nT)$, can be combined to reconstruct $x(t)$.

- From figure, it is clear that larger-than-necessary values of f_s (smaller values of T), called *oversampling*, have no effect on the outcome of the reconstruction and have the benefit of leaving room for a *transition band* in which $H(f)$ is free to take intermediate values. Undersampling, which causes aliasing, is not in general a reversible operation.
- Theoretically, the interpolation formula can be implemented as a low pass filter, whose impulse response is sinc(t/T) and whose input is $\sum_{n=-\infty}^{\infty} x(nT) \cdot \delta(t - nT)$, which is a Dirac comb function modulated by the signal samples. Practical digital-to-analog converters (DAC) implement an approximation like the zero-order hold. In that case, oversampling can reduce the approximation error.

Application to Multivariable Signals and Images

Figure: *Subsampled image showing a Moiré pattern*

The sampling theorem is usually formulated for functions of a single variable. Consequently, the theorem is directly applicable to time-dependent signals and is normally formulated in that context. However, the sampling theorem can be extended in a straightforward way to functions of arbitrarily many variables. Grayscale images, for example, are often represented as two-dimensional arrays (or matrices) of real numbers representing the relative intensities of pixels (picture elements) located at the intersections of row and column sample locations.

As a result, images require two independent variables, or indices, to specify each pixel uniquely — one for the row, and one for the column.

Colour images typically consist of a composite of three separate grayscale images, one to represent each of the three primary colours — red, green, and blue, or *RGB* for short. Other colourspaces using 3-vectors for colours include HSV, LAB, XYZ, etc. Some colourspaces such as cyan, magenta, yellow, and black (CMYK) may represent colour by four dimensions. All of these are treated as vector-valued functions over a two-dimensional sampled domain.

Similar to one-dimensional discrete-time signals, images can also suffer from aliasing if the sampling resolution, or pixel density, is inadequate. For example, a digital photograph of a striped shirt with high frequencies (in other words, the distance between the stripes is small), can cause aliasing of the shirt when it is sampled by the camera's image sensor. The aliasing appears as a moiré pattern. The "solution" to higher sampling in the spatial domain for this case would be to move closer to the shirt, use a higher resolution sensor, or to optically blur the image before acquiring it with the sensor.

Another example is shown to the right in the brick patterns. The top image shows the effects when the sampling theorem's condition is not satisfied. When software rescales an image (the same process that creates the thumbnail shown in the lower image) it, in effect, runs the image through a low-pass filter first and then downsamples the image to result in a smaller image that does not exhibit the moiré pattern. The top image is what happens when the image is downsampled without low-pass filtering: aliasing results.

The application of the sampling theorem to images should be made with care. For example, the sampling process in any standard image sensor (CCD or CMOS camera) is relatively far from the ideal sampling which would measure the image intensity at a single point. Instead these devices have a relatively large sensor area at each sample point in order to obtain sufficient amount of light. In other words, any detector has a finite-width point spread function. The analog optical image intensity function which is sampled by the sensor device is not in general bandlimited, and the non-ideal sampling is itself a useful type of low-pass filter, though not always sufficient to remove enough high frequencies to sufficiently reduce aliasing.

When the area of the sampling spot (the size of the pixel sensor) is not large enough to provide sufficient spatial anti-aliasing, a separate anti-aliasing filter (optical low-pass filter) is typically included in a camera system to further blur the optical image. Despite images having these problems in relation to the sampling theorem, the theorem can be used to describe the basics of down and up sampling of images.

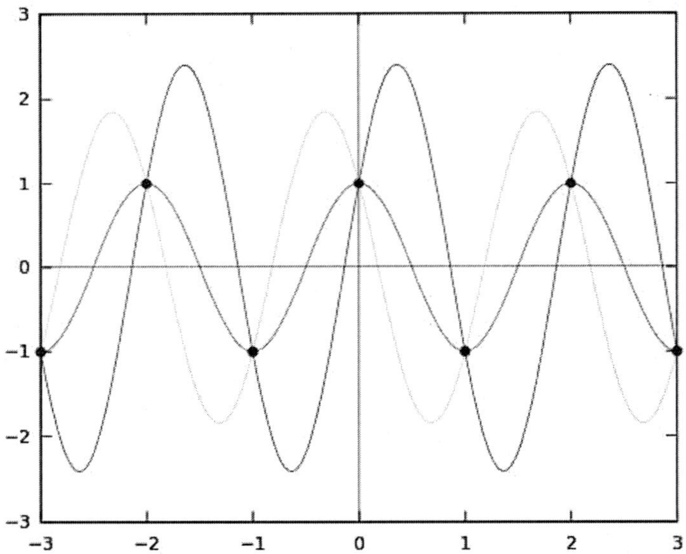

***Figure:** A family of sinusoids at the critical frequency, all having the same sample sequences of alternating +1 and −1. That is, they all are aliases of each other, even though their frequency is not above half the sample rate.*

Critical Frequency

To illustrate the necessity of $f_s > 2B$, consider the family of sinusoids generated by different values of θ in this formula:

$$x(t) = \frac{\cos(2\pi Bt + \theta)}{\cos(\theta)} = \cos(2\pi Bt) - \sin(2\pi Bt)\tan(\theta), \quad -\pi/2 < \theta < \pi/2.$$

With $f_s = 2B$ or equivalently $T = 1/(2B)$, the samples are given by:

$$x(nT) = \cos(\pi n) - \underbrace{\sin(\pi n)}_{0}\tan(\theta) = (-1)^n$$

regardless of the value of θ. That sort of ambiguity is the reason for the *strict* inequality of the sampling theorem's condition.

Sampling of Non-baseband Signals

A similar result is true if the band does not start at zero frequency but at some higher value, and can be proved by a linear translation (corresponding physically to single-sideband modulation) of the zero-frequency case. In this case the elementary pulse is obtained from $\sin(x)/x$ by single-side-band modulation.

That is, a sufficient no-loss condition for sampling signals that do not have baseband components exists that involves the *width* of the non-zero frequency interval as opposed to its highest frequency component.

A bandpass condition is that $X(f) = 0$, for all nonnegative f outside the open band of frequencies:

$$\left(\frac{N}{2}f_s, \frac{N+1}{2}f_s\right),$$

for some nonnegative integer N. This formulation includes the normal baseband condition as the case N=0.

The corresponding interpolation function is the impulse response of an ideal brick-wall bandpass filter (as opposed to the ideal brick-wall lowpass filter used above) with cutoffs at the upper and lower edges of the specified band, which is the difference between a pair of lowpass impulse responses:

$$(N+1)\operatorname{sinc}\left(\frac{(N+1)t}{T}\right) - N\operatorname{sinc}\left(\frac{Nt}{T}\right).$$

Other generalizations, for example to signals occupying multiple non-contiguous bands, are possible as well. Even the most generalized form of the sampling theorem does not have a provably true converse. That is, one cannot conclude that information is necessarily lost just because the conditions of the sampling theorem are not satisfied; from an engineering perspective, however, it is generally safe to assume that if the sampling theorem is not satisfied then information will most likely be lost.

Nonuniform Sampling

The sampling theory of Shannon can be generalized for the case of nonuniform samples, that is, samples not taken equally spaced in time. The Shannon sampling theory for non-uniform sampling states that a band-limited signal can be perfectly reconstructed from its samples if the average sampling rate satisfies the Nyquist condition. Therefore, although uniformly spaced samples may result in easier reconstruction algorithms, it is not a necessary condition for perfect reconstruction.

The general theory for non-baseband and nonuniform samples was developed in 1967 by Landau. He proved that, to paraphrase roughly, the average sampling rate (uniform or otherwise) must be twice the *occupied* bandwidth of the signal, assuming it is *a priori* known what portion of the spectrum was occupied. In the late 1990s, this work was partially extended to cover signals of when the amount of occupied bandwidth was known, but the actual occupied portion of the spectrum was unknown. In the 2000s, a complete theory was developed using compressed sensing. In particular, the theory, using signal processing language, is described in this 2009 paper. They show, among other things, that if the frequency locations are unknown, then it is necessary to sample at least at twice the Nyquist criteria; in other words, you must pay at least a factor of 2 for not knowing the location of the spectrum. Note that minimum sampling requirements do not necessarily guarantee stability.

Beyond Nyquist

The Nyquist–Shannon sampling theorem provides a sufficient condition for the sampling and reconstruction of a band-limited signal. When reconstruction is done via the Whittaker–Shannon interpolation formula, the Nyquist criterion is also a necessary condition to avoid aliasing, in the sense that if samples are taken at a slower rate than twice the band limit, then there are some signals that will not be correctly reconstructed. However, if further restrictions are imposed on the signal, then the Nyquist criterion may no longer be a necessary condition.

A non-trivial example of exploiting extra assumptions about the signal is given by the recent field of compressed sensing, which allows for full reconstruction with a sub-Nyquist sampling rate. Specifically, this applies to signals that are sparse (or compressible) in some domain.

As an example, compressed sensing deals with signals that may have a low over-all bandwidth (say, the *effective* bandwidth *EB*), but the frequency locations are unknown, rather than all together in a single band, so that the passband technique doesn't apply.

In other words, the frequency spectrum is sparse. Traditionally, the necessary sampling rate is thus $2B$. Using compressed sensing techniques, the signal could be perfectly reconstructed if it is sampled at a rate slightly greater than the $2EB$. The downside of this approach is that reconstruction is no longer given by a formula, but instead by the solution to a convex optimization programme which requires well-studied but nonlinear methods.

Historical Background

The sampling theorem was implied by the work of Harry Nyquist in 1928 ("Certain topics in telegraph transmission theory"), in which he showed that up to $2B$ independent pulse samples could be sent through a system of bandwidth B; but he did not explicitly consider the problem of sampling and reconstruction of continuous signals. About the same time, Karl Küpfmüller showed a similar result, and discussed the sinc-function impulse response of a band-limiting filter, via its integral, the step response *Integralsinus*; this bandlimiting and reconstruction filter that is so central to the sampling theorem is sometimes referred to as a *Küpfmüller filter* (but seldom so in English).

The sampling theorem, essentially a dual of Nyquist's result, was proved by Claude E. Shannon in 1949 ("Communication in the presence of noise"). V. A. Kotelnikov published similar results in 1933 ("On the transmission capacity of the 'ether' and of cables in electrical communications", translation from the Russian), as did the mathematician E. T. Whittaker in 1915 ("Expansions of the Interpolation-Theory", "Theorie der Kardinalfunktionen"), J. M. Whittaker in 1935 ("Interpolatory function theory"), and Gabor in 1946 ("Theory of communication").

Other Discoverers

Others who have independently discovered or played roles in the development of the sampling theorem have been discussed in several historical articles, for example by Jerri and by Lüke. For example, Lüke points out that H. Raabe, an assistant to Küpfmüller, proved the theorem in his 1939 Ph.D. dissertation; the term *Raabe condition* came to be associated with the criterion for unambiguous representation (sampling rate greater than twice the bandwidth).

Meijering mentions several other discoverers and names in a paragraph and pair of footnotes:

As pointed out by Higgins, the sampling theorem should really be considered in two parts, as done above: the first stating the fact that a bandlimited function is completely determined by its samples, the second describing how to reconstruct the function using its samples. Both parts of the sampling theorem were given in a somewhat different form by J. M. Whittaker and before him also by Ogura. They were probably not aware of the fact that the first part of the theorem had been stated as early as 1897 by Borel. As we have seen, Borel also used around that time what became known as the cardinal series. However, he appears not to have made the link. In later years it became known that the sampling theorem had been presented before Shannon to the Russian communication community by Kotel'nikov. In more implicit, verbal form, it had also been described in the German literature by Raabe. Several authors have mentioned that Someya introduced the theorem in the Japanese literature parallel to Shannon. In the English literature, Weston introduced it independently of Shannon around the same time.

Several authors, following Black, have claimed that this first part of the sampling theorem was stated even earlier by Cauchy, in a paper published in 1841. However, the paper of Cauchy does not contain such a statement, as has been pointed out by Higgins.

As a consequence of the discovery of the several independent introductions of the sampling theorem, people started to refer to the theorem by including the names of the aforementioned authors, resulting in such catchphrases as "the Whittaker-

Kotel'nikov-Shannon (WKS) sampling theorem" or even "the Whittaker-Kotel'nikov-Raabe-Shannon-Someya sampling theorem". To avoid confusion, perhaps the best thing to do is to refer to it as the sampling theorem, "rather than trying to find a title that does justice to all claimants".

Why Nyquist?

Exactly how, when, or why Harry Nyquist had his name attached to the sampling theorem remains obscure. The term *Nyquist Sampling Theorem* (capitalized thus) appeared as early as 1959 in a book from his former employer, Bell Labs, and appeared again in 1963, and not capitalized in 1965. It had been called the *Shannon Sampling Theorem* as early as 1954, but also just *the sampling theorem* by several other books in the early 1950s.

In 1958, Blackman and Tukey cited Nyquist's 1928 paper as a reference for *the sampling theorem of information theory*, even though that paper does not treat sampling and reconstruction of continuous signals as others did. Their glossary of terms includes these entries:

Sampling theorem (of information theory): Nyquist's result that equi-spaced data, with two or more points per cycle of highest frequency, allows reconstruction of band-limited functions.

Cardinal theorem (of interpolation theory): A precise statement of the conditions under which values given at a doubly infinite set of equally spaced points can be interpolated to yield a continuous band-limited function with the aid of the function

$$\frac{\sin(x - x_i)}{x - x_i}.$$

Exactly what "Nyquist's result" they are referring to remains mysterious.

When Shannon stated and proved the sampling theorem in his 1949 paper, according to Meijering "he referred to the critical sampling interval $T = 1/(2W)$ as the *Nyquist interval* corresponding to the band W, in recognition of Nyquist's discovery of the fundamental importance of this interval in connection with telegraphy." This explains Nyquist's name on the critical interval, but not on the theorem.

Similarly, Nyquist's name was attached to *Nyquist rate* in 1953 by Harold S. Black:

> *"If the essential frequency range is limited to* B *cycles per second,* 2B *was given by Nyquist as the maximum number of code elements per second that could be unambiguously resolved, assuming the peak interference is less half a quantum step. This rate is generally referred to as signaling at the Nyquist rate and* 1/(2B) *has been termed a* Nyquist interval.*" (bold added for emphasis; italics as in the original)*

According to the OED, this may be the origin of the term *Nyquist rate.* In Black's usage, it is not a sampling rate, but a signaling rate.

Use in Directional Statistics

In directional statistics, the Dirac comb of period 2π is equivalent to a wrapped Dirac delta function, and is the analog of the Dirac delta function in linear statistics.

In linear statistics, the random variable (x) is usually distributed over the real number line, or some subset thereof, and the probability density of x is a function whose domain is the set real numbers, and whose integral from $-\infty$ to $+\infty$ is unity. In directional statistics, the random variable (θ) is distributed over the unit circle and the probability density of θ is a function whose domain is some interval of the real numbers of length 2π and whose integral over that interval is unity. Just as the integral of the product of a Dirac delta function with an arbitrary function over the real number line yields the value of that function at zero, so the integral of the product of a Dirac comb of period 2π with an arbitrary function of period 2π over the unit circle yields the value of that function at zero.

Sokhotski–Plemelj theorem

The Sokhotski–Plemelj theorem, important in quantum mechanics, relates the delta function to the distribution p.v.1/x, the Cauchy principal value of the function $1/x$, defined by

$$\left\langle \text{p.v.}\frac{1}{x}, \phi \right\rangle = \lim_{\varepsilon \to 0^+} \int_{|x|>\varepsilon} \frac{\phi(x)}{x} dx.$$

Sokhatsky's formula states that

$$\lim_{\varepsilon \to 0^+} \frac{1}{x \pm i\varepsilon} = \text{p.v.}\frac{1}{x} \mp i\pi\delta(x),$$

Here the limit is understood in the distribution sense, that for all compactly supported smooth functions *f*,

$$\lim_{\varepsilon \to 0^+} \int_{-\infty}^{\infty} \frac{f(x)}{x \pm i\varepsilon} dx = \mp i\pi f(0) + \lim_{\varepsilon \to 0^+} \int_{|x|>\varepsilon} \frac{f(x)}{x} dx.$$

Relationship to the Kronecker Delta

The Kronecker delta δ_{ij} is the quantity defined by

$$\delta_{ij} = \begin{cases} 1 & i = j \\ 0 & i \neq j \end{cases}$$

for all integers *i*, *j*. This function then satisfies the following analog of the sifting property: if $(a_i)_{i \in Z}$ is any doubly infinite sequence, then

$$\sum_{i=-\infty}^{\infty} a_i \delta_{ik} = a_k.$$

Similarly, for any real or complex valued continuous function *f* on R, the Dirac delta satisfies the sifting property

$$\int_{-\infty}^{\infty} f(x)\delta(x - x_0)dx = f(x_0).$$

This exhibits the Kronecker delta function as a discrete analog of the Dirac delta function.

Applications to Probability Theory

In probability theory and statistics, the Dirac delta function is often used to represent a discrete distribution, or a partially discrete, partially continuous distribution, using a probability density function (which is normally used to represent fully continuous distributions). For example, the probability density function *f*(*x*) of a discrete distribution consisting of points x = $\{x_1, \ldots, x_n\}$, with corresponding probabilities $p_1, \ldots, p_n$, can be written as

$$f(x) = \sum_{i=1}^{n} p_i \delta(x - x_i).$$

As another example, consider a distribution which 6/10 of the time returns a standard normal distribution, and 4/10 of the time returns exactly the value 3.5 (i.e. a partly continuous, partly discrete mixture distribution). The density function of this distribution can be written as

$$f(x) = 0.6\frac{1}{\sqrt{2\pi}}e^{-\frac{x^2}{2}} + 0.4\delta(x-3.5).$$

The delta function is also used in a completely different way to represent the local time of a diffusion process (like Brownian motion). The local time of a stochastic process $B(t)$ is given by

$$\ell(x,t) = \int_0^t \delta(x - B(s))ds$$

and represents the amount of time that the process spends at the point x in the range of the process. More precisely, in one dimension this integral can be written

$$\ell(x,t) = \lim_{\varepsilon\to 0^+}\frac{1}{2\varepsilon}\int_0^t 1_{[x-\varepsilon,x+\varepsilon]}(B(s))ds$$

where $1_{[x-\varepsilon,\, x+\varepsilon]}$ is the indicator function of the interval $[x-\varepsilon, x+\varepsilon]$.

Application to Quantum Mechanics

We give an example of how the delta function is expedient in quantum mechanics. The wave function of a particle gives the probability amplitude of finding a particle within a given region of space. Wave functions are assumed to be elements of the Hilbert space L^2 of square-integrable functions, and the total probability of finding a particle within a given interval is the integral of the magnitude of the wave function squared over the interval. A set $\{\varphi_n\}$ of wave functions is orthonormal if they are normalized by

$$\langle \phi_n \mid \phi_m \rangle = \delta_{nm}$$

where δ here refers to the Kronecker delta. A set of orthonormal wave functions is complete in the space of square-integrable functions if any wave function ψ can be expressed as a combination of the φ_n:

$$\psi = \sum c_n \phi_n,$$

with $c_n = \langle \phi_n \mid \psi \rangle$. Complete orthonormal systems of wave functions appear naturally as the eigenfunctions of the Hamiltonian (of a bound system) in quantum mechanics that measures the energy levels, which are called the eigenvalues. The set of eigenvalues, in this case, is known as the spectrum of the Hamiltonian. In bra-ket notation, as above, this equality implies the resolution of the identity:

$$I = \sum |\phi_n\rangle\langle\phi_n|.$$

Here the eigenvalues are assumed to be discrete, but the set of eigenvalues of an observable may be continuous rather than discrete. An example is the position observable, $Q\psi(x) = x\psi(x)$. The spectrum of the position (in one dimension) is the entire real line, and is called a continuous spectrum. However, unlike the Hamiltonian, the position operator lacks proper eigenfunctions. The conventional way to overcome this shortcoming is to widen the class of available functions by allowing distributions as well: that is, to replace the Hilbert space of quantum mechanics by an appropriate rigged Hilbert space. In this context, the position operator has a complete set of eigen-distributions, labelled by the points y of the real line, given by

$$\phi_y(x) = \delta(x - y).$$

The eigenfunctions of position are denoted by $\phi_y = |y\rangle$ in Dirac notation, and are known as position eigenstates.

Similar considerations apply to the eigenstates of the momentum operator, or indeed any other self-adjoint unbounded operator P on the Hilbert space, provided the spectrum of P is continuous and there are no degenerate eigenvalues. In that case, there is a set Ω of real numbers (the spectrum), and a collection φ_y of distributions indexed by the elements of Ω, such that

$$P\phi_y = y\phi_y.$$

That is, φ_y are the eigenvectors of P. If the eigenvectors are normalized so that

$$\langle \phi_y, \phi_{y'} \rangle = \delta(y - y')$$

in the distribution sense, then for any test function ψ,

$$\psi(x) = \int_{\Omega} c(y)\phi_y(x)dy$$

where

$$c(y) = \langle \psi, \phi_y \rangle.$$

That is, as in the discrete case, there is a resolution of the identity

$$I = \int_{\Omega} |\phi_y\rangle\langle\phi_y| dy$$

where the operator-valued integral is again understood in the weak sense. If the spectrum of P has both continuous and discrete parts, then the resolution of the identity involves a summation over the discrete spectrum *and* an integral over the continuous spectrum.

The delta function also has many more specialized applications in quantum mechanics, such as the delta potential models for a single and double potential well.

Application to Structural Mechanics

The delta function can be used in structural mechanics to describe transient loads or point loads acting on structures. The governing equation of a simple mass–spring system excited by a sudden force impulse I at time $t = 0$ can be written

$$m\frac{\mathrm{d}^2\xi}{\mathrm{d}t^2} + k\xi = I\delta(t),$$

where m is the mass, ∂ the deflection and k the spring constant.

As another example, the equation governing the static deflection of a slender beam is, according to Euler-Bernoulli theory,

$$EI\frac{\mathrm{d}^4 w}{\mathrm{d}x^4} = q(x),$$

where EI is the bending stiffness of the beam, w the deflection, x the spatial coordinate and $q(x)$ the load distribution. If a beam is loaded by a point force F at $x = x_0$, the load distribution is written

$$q(x) = F\delta(x - x_0).$$

As integration of the delta function results in the Heaviside step function, it follows that the static deflection of a slender beam subject to multiple point loads is described by a set of piecewise polynomials.

Also a point moment acting on a beam can be described by delta functions. Consider two opposing point forces F at a distance d apart. They then produce a moment $M = Fd$ acting on the beam. Now, let the distance d approach the limit zero, while M is kept constant. The load distribution, assuming a clockwise moment acting at $x = 0$, is written

$$\begin{aligned} q(x) &= \lim_{d\to 0}\left(F\delta(x) - F\delta(x-d)\right) \\ &= \lim_{d\to 0}\left(\frac{M}{d}\delta(x) - \frac{M}{d}\delta(x-d)\right) \\ &= M\lim_{d\to 0}\frac{\delta(x)-\delta(x-d)}{d} \\ &= M\delta'(x). \end{aligned}$$

Point moments can thus be represented by the derivative of the delta function. Integration of the beam equation again results in piecewise polynomial deflection.

Heaviside Step Function

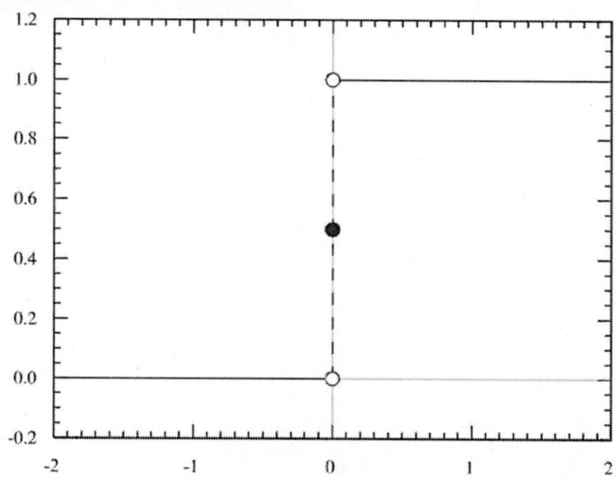

Figure: *The Heaviside step function, using the half-maximum convention*

The Heaviside step function, or the unit step function, usually denoted by H (but sometimes u or θ), is a discontinuous function whose value is zero for negative argument and one for positive argument. It seldom matters what value is used for $H(0)$, since H is mostly used as a distribution. Some common choices can be seen below.

The function is used in the mathematics of control theory and signal processing to represent a signal that switches on at a specified time and stays switched on indefinitely. It is also used in structural mechanics together with the Dirac delta function to describe different types of structural loads. It was named after the English polymath Oliver Heaviside.

It is the cumulative distribution function of a random variable which is almost surely 0.

The Heaviside function is the integral of the Dirac delta function: $H2 = \delta$. This is sometimes written as

$$H(x) = \int_{-\infty}^{x} \delta(s)\mathrm{d}s$$

although this expansion may not hold (or even make sense) for $x = 0$, depending on which formalism one uses to give meaning to integrals involving δ.

Discrete Form

An alternative form of the unit step, as a function of a discrete variable n:

$$H[n] = \begin{cases} 0, & n < 0, \\ 1, & n \geq 0, \end{cases}$$

where n is an integer. Unlike the usual (not discrete) case, the definition of $H[0]$ is significant.

The discrete-time unit impulse is the first difference of the discrete-time step

$$\delta[n] = H[n] - H[n-1].$$

This function is the cumulative summation of the Kronecker delta:

$$H[n] = \sum_{k=-\infty}^{n} \delta[k]$$

where

$$\delta[k] = \delta_{k,0}$$

is the discrete unit impulse function.

Analytic Approximations

For a smooth approximation to the step function, one can use the logistic function

$$H(x) \approx \frac{1}{2} + \frac{1}{2}\tanh(kx) = \frac{1}{1+e^{-2kx}},$$

where a larger k corresponds to a sharper transition at $x = 0$. If we take $H(0) = ½$, equality holds in the limit:

$$H(x) = \lim_{k\to\infty} \frac{1}{2}(1 + \tanh kx) = \lim_{k\to\infty} \frac{1}{1+e^{-2kx}}.$$

There are many other smooth, analytic approximations to the step function. Among the possibilities are:

$$H(x) = \lim_{k\to\infty}\left(\frac{1}{2} + \frac{1}{\pi}\arctan(kx)\right)$$

$$H(x) = \lim_{k\to\infty}\left(\frac{1}{2} + \frac{1}{2}\operatorname{erf}(kx)\right)$$

These limits hold pointwise and in the sense of distributions. In general, however, pointwise convergence need not imply distributional convergence, and vice-versa distributional convergence need not imply pointwise convergence.

In general, any cumulative distribution function (c.d.f.) of a continuous probability distribution that is peaked around zero and has a parameter that controls for variance can serve as an approximation, in the limit as the variance approaches zero. For example, all three of the above approximations are c.d.f.s of common probability distributions: The logistic, Cauchy and normal distributions, respectively.

Integral Representations

Often an integral representation of the Heaviside step function is useful:

$$H(x) = \lim_{\epsilon\to 0^+} -\frac{1}{2\pi i}\int_{-\infty}^{\infty} \frac{1}{\tau + i\epsilon} e^{-ix\tau} d\tau = \lim_{\epsilon\to 0^+} \frac{1}{2\pi i}\int_{-\infty}^{\infty} \frac{1}{\tau - i\epsilon} e^{ix\tau} d\tau.$$

Zero Argument

Since H is usually used in integration, and the value of a function at a single point does not affect its integral, it rarely matters what particular value is chosen of $H(0)$. Indeed when H is considered as a distribution or an element of L^∞ it does not even make sense to talk of a value at zero, since such objects are only defined almost everywhere. If using some analytic approximation (as in the examples above) then often whatever happens to be the relevant limit at zero is used.

There exist various reasons for choosing a particular value.

- $H(0) = \frac{1}{2}$ is often used since the graph then has rotational symmetry; put another way, $H-\frac{1}{2}$ is then an odd function. In this case the following relation with the sign function holds for all x:

  $$H(x) = \tfrac{1}{2}(1 + \operatorname{sgn}(x)).$$

- $H(0) = 1$ is used when H needs to be right-continuous. For instance cumulative distribution functions are usually taken to be right continuous, as are functions integrated against in Lebesgue–Stieltjes integration. In this case H is the indicator function of a closed semi-infinite interval:

  $$H(x) = 1_{[0,\infty)}(x).$$

- $H(0) = 0$ is used when H needs to be left-continuous. In this case H is an indicator function of an open semi-infinite interval:

  $$H(x) = 1_{(0,\infty)}(x).$$

Antiderivative and Derivative

The ramp function is the antiderivative of the Heaviside step function: $R(x) := \int_{-\infty}^{x} H(\xi)d\xi = xH(x)$.

The distributional derivative of the Heaviside step function is the Dirac delta function: $\frac{dH(x)}{dx} = \delta(x)$

Generalization

The indicator function of a domain D is the function that is equal to one inside D and zero otherwise. The Heaviside function is the indicator function of $[0,\infty)$. In higher dimensions, the indicator function of a domain still makes sense, and thus makes a natural generalization of the Heaviside step function.

The inward normal derivative of the indicator gives a *surface* Dirac delta function:

$$-\int_{\mathbb{R}^n} g(\mathrm{x})\mathrm{n}_x \cdot \nabla_x 1_{\mathrm{x}\in D}\, d\mathrm{x} = \oint_{\partial D} g(\beta)\, d\beta,$$

where n is the outward normal.

Fourier Transform

The Fourier transform of the Heaviside step function is a distribution. Using one choice of constants for the definition of the Fourier transform we have

$$\hat{H}(s) = \lim_{N\to\infty}\int_{-N}^{N} \mathrm{e}^{-2\pi i x s} H(x)\mathrm{d}x = \frac{1}{2}\left(\delta(s) - \frac{i}{\pi}\mathrm{p.v.}\frac{1}{s}\right).$$

Here $\mathrm{p.v.}\frac{1}{s}$ is the distribution that takes a test function ϕ to the Cauchy principal value of $\int_{-\infty}^{\infty} \varphi(s)/s\,ds$. The limit appearing in the integral is also taken in the sense of (tempered) distributions.

Hyperfunction Representation

This can be represented as a hyperfunction as $H(x) = \left(\frac{1}{2\pi i}\log(z), \frac{1}{2\pi i}\log(z) - 1\right)$.

Hilbert Transform

In mathematics and in signal processing, the Hilbert transform is a linear operator which takes a function, $u(t)$, and produces a function, $H(u)(t)$, with the same domain. The Hilbert transform is named after David Hilbert, who first introduced the operator in order to solve a special case of the Riemann–Hilbert problem for holomorphic functions. It is a basic tool in Fourier analysis, and provides a concrete means for realizing the harmonic conjugate of a given function or Fourier series. Furthermore, in harmonic analysis, it is an example of a singular integral operator, and of a Fourier multiplier. The Hilbert transform is also important in the field of signal processing where it is used to derive the analytic representation of a signal $u(t)$.

The Hilbert transform was originally defined for periodic functions, or equivalently for functions on the circle, in which case it is given by convolution with the Hilbert kernel. More commonly, however, the Hilbert transform refers to a convolution with the Cauchy kernel, for functions defined on the real line R (the boundary of the upper half-plane). The Hilbert transform is closely related to the Paley–Wiener theorem, another result relating holomorphic functions in the upper half-plane and Fourier transforms of functions on the real line.

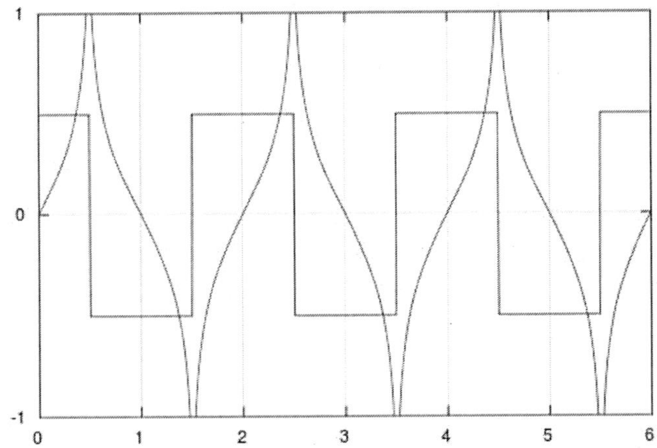

Figure: *The Hilbert transform, in red, of a square wave, in blue*

The Hilbert transform of u can be thought of as the convolution of $u(t)$ with the function $h(t) = 1/(\pi t)$. Because $h(t)$ is not integrable the integrals defining the convolution do not converge. Instead, the Hilbert transform is defined using the Cauchy principal value (denoted here by p.v.) Explicitly, the Hilbert transform of a function (or signal) $u(t)$ is given by

$$H(u)(t) = \text{p.v.}\int_{-\infty}^{\infty} u(\tau)h(t-\tau)d\tau = \frac{1}{\pi}\,\text{p.v.}\int_{-\infty}^{\infty} \frac{u(\tau)}{t-\tau}d\tau,$$

provided this integral exists as a principal value. This is precisely the convolution of u with the tempered distribution p.v. $1/\pi t$ (due to Schwartz (1950). Alternatively, by changing variables, the principal value integral can be written explicitly (Zygmund 1968, §XVI.1) as

$$H(u)(t) = -\frac{1}{\pi}\lim_{\varepsilon\downarrow 0}\int_{\varepsilon}^{\infty}\frac{u(t+\tau)-u(t-\tau)}{\tau}d\tau.$$

When the Hilbert transform is applied twice in succession to a function u, the result is negative u:

$$H(H(u))(t) = -u(t),$$

provided the integrals defining both iterations converge in a suitable sense. In particular, the inverse transform is $-H$. This fact can most easily be seen by considering the effect of the Hilbert transform on the Fourier transform of $u(t)$.

For an analytic function in upper half-plane the Hilbert transform describes the relationship between the real part and the imaginary part of the boundary values. That is, if $f(z)$ is analytic in the plane Im $z > 0$ and $u(t)$ = Re $f(t + 0.i)$ then Im $f(t + 0 \cdot i) = H(u)(t)$ up to an additive constant, provided this Hilbert transform exists.

Notation

In signal processing the Hilbert transform of $u(t)$ is commonly denoted by $\hat{u}(t)$. However, in mathematics, this notation is already extensively used to denote the Fourier transform of $u(t)$. Occasionally, the Hilbert transform may be denoted by $\hat{u}(t)$. . Furthermore, many sources define the Hilbert transform as the negative of the one defined here.

History

The Hilbert transform arose in Hilbert's 1905 work on a problem posed by Riemann concerning analytic functions (Kress (1989); Bitsadze (2001)) which has come to be known as the Riemann–Hilbert problem. Hilbert's work was mainly concerned with the Hilbert transform for functions defined on the circle (Khvedelidze 2001; Hilbert 1953). Some of his earlier work related to the Discrete Hilbert Transform dates back to lectures he gave in Göttingen. The results were later published by Hermann Weyl in his dissertation (Hardy, Littlewood & Polya 1952, §9.1). Schur improved Hilbert's results about the discrete Hilbert transform and extended them to the integral case (Hardy, Littlewood & Polya 1952, §9.2). These results were restricted to the spaces L^2 and ℓ^2. In 1928, Marcel Riesz proved that the Hilbert transform can be defined for u in $L^p(\mathrm{R})$ for

$1 \leq p \leq \infty$ that the Hilbert transform is a bounded operator on $L^p(\mathbb{R})$ for the same range of p, and that similar results hold for the Hilbert transform on the circle as well as the discrete Hilbert transform (Riesz 1928). The Hilbert transform was a motivating example for Antoni Zygmund and Alberto Calderón during their study of singular integrals (Calderón & Zygmund 1952). Their investigations have played a fundamental role in modern harmonic analysis. Various generalizations of the Hilbert transform, such as the bilinear and trilinear Hilbert transforms are still active areas of research today.

Relationship with the Fourier Transform

The Hilbert transform is a multiplier operator (Duoandikoetxea 2000, Chapter 3). The symbol of H is $\sigma_H(\omega) = -i\,\mathrm{sgn}(\omega)$ where sgn is the signum function. Therefore:

$$\mathcal{F}(H(u))(\omega) = (-i\,\mathrm{sgn}(\omega)) \cdot \mathcal{F}(u)(\omega)$$

where $\mathcal{F}$ denotes the Fourier transform. Since $\mathrm{sgn}(x) = \mathrm{sgn}(2\pi x)$, it follows that this result applies to the three common definitions of $\mathcal{F}$.

By Euler's formula,

$$\sigma_H(\omega) = \begin{cases} i = e^{+i\pi/2}, & \text{for } \omega < 0 \\ 0, & \text{for } \omega = 0 \\ -i = e^{-i\pi/2}, & \text{for } \omega > 0. \end{cases}$$

Therefore $H(u)(t)$ has the effect of shifting the phase of the negative frequency components of $u(t)$ by +90° ($\pi/2$ radians) and the phase of the positive frequency components by –90°. And $i \cdot H(u)(t)$ has the effect of restoring the positive frequency components while shifting the negative frequency ones an additional +90°, resulting in their negation.

When the Hilbert transform is applied twice, the phase of the negative and positive frequency components of $u(t)$ are respectively shifted by +180° and –180°, which are equivalent amounts. The signal is negated, i.e., $H(H(u)) = -u$, because:

$$(\sigma_H(\omega))^2 = e^{\pm i\pi} = -1 \qquad \text{for } \omega \neq 0.$$

Domain of Definition

It is by no means obvious that the Hilbert transform is well-defined at all, as the improper integral defining it must converge in a suitable sense. However, the Hilbert transform is well-defined for a broad class of functions, namely those in $L^p(\mathrm{R})$ for $1 < p < \infty$.

More precisely, if u is in $L^p(\mathrm{R})$ for $1<p<\infty$, then the limit defining the improper integral

$$H(u)(t) = -\frac{1}{\pi}\lim_{\epsilon\downarrow 0}\int_{\epsilon}^{\infty}\frac{u(t+\tau)-u(t-\tau)}{\tau}d\tau$$

exists for almost every t. The limit function is also in $L^p(\mathrm{R})$, and is in fact the limit in the mean of the improper integral as well. That is,

$$-\frac{1}{\pi}\int_{\epsilon}^{\infty}\frac{u(t+\tau)-u(t-\tau)}{\tau}d\tau \to H(u)(t)$$

as $\varepsilon\to 0$ in the L^p-norm, as well as pointwise almost everywhere, by the Titchmarsh theorem (Titchmarsh 1948, Chapter 5).

In the case $p=1$, the Hilbert transform still converges pointwise almost everywhere, but may fail to be itself integrable even locally (Titchmarsh 1948, §5.14). In particular, convergence in the mean does not in general happen in this case. The Hilbert transform of an L^1 function does converge, however, in L^1-weak, and the Hilbert transform is a bounded operator from L^1 to $L^{1,w}$ (Stein & Weiss 1971, Lemma V.2.8). (In particular, since the Hilbert transform is also a multiplier operator on L^2, Marcinkiewicz interpolation and a duality argument furnishes an alternative proof that H is bounded on L^p.)

Properties

Boundedness: If $1<p<\infty$, then the Hilbert transform on $L^p(\mathrm{R})$ is a bounded linear operator, meaning that there exists a constant C_p such that

$$\| Hu \|_p \le C_p \| u \|_p$$

for all $u \in L^p(\mathrm{R})$. This theorem is due to Riesz (1928, VII). The best constant C_p is given by

$$C_p = \begin{cases} \tan\dfrac{\pi}{2p} & \text{for } 1 < p \le 2 \\ \cot\dfrac{\pi}{2p} & \text{for } 2 < p < \infty. \end{cases}$$

This result is due to (Pichorides 1972). The same best constants hold for the periodic Hilbert transform.

The boundedness of the Hilbert transform implies the $L^p(\mathrm{R})$ convergence of the symmetric partial sum operator

$$S_R f = \int_{-R}^{R} \hat{f}\xi e^{2\pi i x\xi}\, d\xi$$

to f in $L^p(\mathrm{R})$.

Anti-self Adjointness

The Hilbert transform is an anti-self adjoint operator relative to the duality pairing between $L^p(\mathrm{R})$ and the dual space $L^q(\mathrm{R})$, where p and q are Hölder conjugates and $1 < p,q < \infty$. Symbolically,

$$\langle Hu,v\rangle = \langle u,-Hv\rangle$$

for $u \in L^p(\mathrm{R})$ and $v \in L^q(\mathrm{R})$ (Titchmarsh 1948, Theorem 102).

Inverse Transform

The Hilbert transform is an anti-involution (Titchmarsh 1948, p. 120), meaning that

$$H(H(u)) = -u$$

provided each transform is well-defined. Since H preserves the space $L^p(\mathrm{R})$, this implies in particular that the Hilbert transform is invertible on $L^p(\mathrm{R})$, and that

$$H^{-1} = -H.$$

Differentiation

Formally, the derivative of the Hilbert transform is the Hilbert transform of the derivative, i.e. these two linear operators commute:

$$H\left(\frac{du}{dt}\right) = \frac{d}{dt} H(u).$$

Iterating this identity,

$$H\left(\frac{d^k u}{dt^k}\right) = \frac{d^k}{dt^k} H(u).$$

This is rigorously true as stated provided u and its first k derivatives belong to $L^p(\mathrm{R})$ (Pandey 1996, §3.3). One can check this easily in the frequency domain, where differentiation becomes multiplication by ω.

Convolutions

The Hilbert transform can formally be realized as a convolution with the tempered distribution (Duistermaat & Kolk 2010, p. 211)

$$h(t) = \text{p.v.}\ \frac{1}{\pi t}.$$

Thus formally,

$$H(u) = h * u.$$

However, *a priori* this may only be defined for u a distribution of compact support. It is possible to work somewhat rigorously with this since compactly supported functions (which are distributions *a fortiori*) are dense in L^p. Alternatively, one may use the fact that $h(t)$ is the distributional derivative of the function $\log|t|/\pi$; to wit

$$H(u)(t) = \frac{d}{dt}\left(\frac{1}{\pi}(u * \log|\cdot|)(t)\right).$$

For most operational purposes the Hilbert transform can be treated as a convolution. For example, in a formal sense, the Hilbert transform of a convolution is the convolution of the Hilbert transform on either factor:

$$H(u * v) = H(u) * v = u * H(v).$$

This is rigorously true if u and v are compactly supported distributions since, in that case,

$$h * (u * v) = (h * u) * v = u * (h * v).$$

By passing to an appropriate limit, it is thus also true if $u \in L^p$ and $v \in L^r$ provided

$$1 < \frac{1}{p} + \frac{1}{r},$$

a theorem due to Titchmarsh (1948, Theorem 104).

Invariance

The Hilbert transform has the following invariance properties on $L^2(\mathbb{R})$.

- It commutes with translations. That is, it commutes with the operators $T_a f(x) = f(x + a)$ for all a in R
- It commutes with positive dilations. That is it commutes with the operators $M_\lambda f(x) = f(\lambda x)$ for all $\lambda > 0$.
- It anticommutes with the reflection $Rf(x) = f(-x)$.

Up to a multiplicative constant, the Hilbert transform is the only bounded operator on L^2 with these properties (Stein 1970, §III.1).

In fact there is a larger group of operators commuting with the Hilbert transform. The group SL(2,R) acts by unitary operators U_g on the space $L^2(\mathrm{R})$ by the formula

$$U_g^{-1} f(x) = (cx+d)^{-1} f\left(\frac{ax+b}{cx+d}\right),\ g = \begin{pmatrix} a & b \\ c & d \end{pmatrix}.$$

This unitary representation is an example of a principal series representation of SL(2,R). In this case it is reducible, splitting as the orthogonal sum of two invariant subspaces, Hardy space $H^2(\mathrm{R})$ and its conjugate. These are the spaces of L^2 boundary values of holomorphic functions on the upper and lower halfplanes. $H^2(\mathrm{R})$ and its conjugate consist of exactly those L^2 functions with Fourier transforms vanishing on the negative and positive parts of the real axis respectively. Since the Hilbert transform is equal to $H = -i\,(2P - I)$, with P being the orthogonal projection from $L^2(\mathrm{R})$ onto $H^2(\mathrm{R})$, it follows that $H^2(\mathrm{R})$ and its orthogonal are eigenspaces of H for the eigenvalues $\pm\, i$. In other words H commutes with the operators U_g. The restrictions of the operators U_g to $H^2(\mathrm{R})$ and its conjugate give irreducible representations of SL(2,R)—the so-called limit of discrete series representations.

Extending the Domain of Definition

Hilbert Transform of Distributions: It is further possible to extend the Hilbert transform to certain spaces of distributions (Pandey 1996, Chapter 3). Since the Hilbert transform commutes with differentiation, and is a bounded operator on L^p, H restricts to give a continuous transform on the inverse limit of Sobolev spaces:

$$\mathcal{D}_{L^p} = \lim_{\substack{\leftarrow \\ n\to\infty}} W^{n,p}(\mathbb{R}).$$

The Hilbert transform can then be defined on the dual space of $\mathcal{D}_{L^p}$, denoted $\mathcal{D}_{L^{p'}}$, consisting of L^p distributions. This is accomplished by the duality pairing: for $u \in \mathcal{D}'_{L^p}$, define $H(u) \in \mathcal{D}'_{L^p}$ by

$$\langle Hu, v\rangle \overset{\text{def}}{=} \langle u, -Hv\rangle$$

for all $v \in \mathcal{D}_{L^p}$.

It is possible to define the Hilbert transform on the space of tempered distributions as well by an approach due to Gel'fand & Shilov (1967), but considerably more care is needed because of the singularity in the integral.

Hilbert Transform of Bounded Functions

The Hilbert transform can be defined for functions in $L^\infty(\mathrm{R})$ as well, but it requires some modifications and caveats. Properly understood, the Hilbert transform maps $L^\infty(\mathrm{R})$ to the Banach space of bounded mean oscillation (BMO) classes.

Interpreted naively, the Hilbert transform of a bounded function is clearly ill-defined. For instance, with $u = \operatorname{sgn}(x)$, the integral defining $H(u)$ diverges almost everywhere to $\pm\infty$. To alleviate such difficulties, the Hilbert transform of an L^∞-function is therefore defined by the following regularized form of the integral

$$H(u)(t) = \text{p.v.}\int_{-\infty}^{\infty} u(\tau)\{h(t-\tau) - h_0(-\tau)\}d\tau$$

where as above $h(x) = 1/\pi x$ and

$$h_0(x) = \begin{cases} 0 & \text{if } |x| < 1 \\ \dfrac{1}{\pi x} & \text{otherwise} \end{cases}$$

The modified transform H agrees with the original transform on functions of compact support by a general result of Calderón & Zygmund (1952). The resulting integral, furthermore, converges pointwise almost everywhere, and with respect to the BMO norm, to a function of bounded mean oscillation.

A deep result of Fefferman (1971) and Fefferman & Stein (1972) is that a function is of bounded mean oscillation if and only if it has the form $f + H(g)$ for some $f, g \in L^\infty(\mathrm{R})$.

Conjugate Functions

The Hilbert transform can be understood in terms of a pair of functions $f(x)$ and $g(x)$ such that the function

$$F(x) = f(x) + ig(x)$$

is the boundary value of a holomorphic function $F(z)$ in the upper half-plane (Titchmarsh 1948, Chapter V). Under these circumstances, if f and g are sufficiently integrable, then one is the Hilbert transform of the other.

Suppose that $f \in L^p(\mathrm{R})$. Then, by the theory of the Poisson integral, f admits a unique harmonic extension into the upper half-plane, and this extension is given by

$$u(x+iy) = u(x,y) = \frac{1}{\pi}\int_{-\infty}^{\infty} f(s)\frac{y}{(x-s)^2+y^2}ds$$

which is the convolution of f with the Poisson kernel

$$P(x,y) = \frac{1}{\pi}\frac{y}{x^2+y^2}.$$

Furthermore, there is a unique harmonic function v defined in the upper half-plane such that $F(z) = u(z) + iv(z)$ is holomorphic and

$$\lim_{y\to\infty} v(x+iy) = 0.$$

This harmonic function is obtained from f by taking a convolution with the conjugate Poisson kernel

$$Q(x,y) = \frac{1}{\pi}\frac{x}{x^2+y^2}.$$

Thus

$$v(x,y) = \frac{1}{\pi}\int_{-\infty}^{\infty} f(s)\frac{x-s}{(x-s)^2+y^2}ds.$$

Indeed, the real and imaginary parts of the Cauchy kernel are

$$\frac{i}{\pi z} = P(x,y) + iQ(x,y),$$

so that $F = u + iv$ is holomorphic by the Cauchy integral theorem.

The function v obtained from u in this way is called the harmonic conjugate of u. The (non-tangential) boundary limit of $v(x,y)$ as $y \to 0$ is the Hilbert transform of f. Thus, succinctly,

$$H(f) = \lim_{y \to 0} Q(-, y) \star f.$$

Titchmarsh's Theorem

A theorem due to Edward Charles Titchmarsh makes precise the relationship between the boundary values of holomorphic functions in the upper half-plane and the Hilbert transform (Titchmarsh 1948, Theorem 95). It gives necessary and sufficient conditions for a complex-valued square-integrable function $F(x)$ on the real line to be the boundary value of a function in the Hardy space $H^2(U)$ of holomorphic functions in the upper half-plane U.

The theorem states that the following conditions for a complex-valued square-integrable function $F: \mathrm{R} \to \mathrm{C}$ are equivalent:

- $F(x)$ is the limit as $z \to x$ of a holomorphic function $F(z)$ in the upper half-plane such that

$$\int_{-\infty}^{\infty} |F(x+iy)|^2 \, dx < K.$$

- $-\mathrm{Im}(F)$ is the Hilbert transform of $\mathrm{Re}(F)$, where $\mathrm{Re}(F)$ and $\mathrm{Im}(F)$ are real-valued functions with $F = \mathrm{Re}(F) + \mathrm{i}\ \mathrm{Im}(F)$.
- The Fourier transform $\mathcal{F}(F)(x)$ vanishes for $x < 0$.

A weaker result is true for functions of class L^p for $p > 1$ (Titchmarsh 1948, Theorem 103). Specifically, if $F(z)$ is a holomorphic function such that

$$\int_{-\infty}^{\infty} |F(x+iy)|^p \, dx < K$$

for all y, then there is a complex-valued function $F(x)$ in $L^p(\mathrm{R})$ such that $F(x + \mathrm{i}y) \to F(x)$ in the L^p norm as $y \to 0$ (as well as holding pointwise almost everywhere). Furthermore,

$$F(x) = f(x) - ig(x)$$

where f is a real-valued function in $L^p(\mathrm{R})$ and g is the Hilbert transform (of class L^p) of f.

This is not true in the case $p = 1$. In fact, the Hilbert transform of an L^1 function f need not converge in the mean to another L^1 function. Nevertheless (Titchmarsh 1948, Theorem

105), the Hilbert transform of f does converge almost everywhere to a finite function g such that

$$\int_{-\infty}^{\infty} \frac{|g(x)|^p}{1+x^2} dx < \infty.$$

This result is directly analogous to one by Andrey Kolmogorov for Hardy functions in the disc (Duren 1970, Theorem 4.2).

Riemann–Hilbert Problem

One form of the Riemann–Hilbert problem seeks to identify pairs of functions F_+ and F_- such that F_+ is holomorphic on the upper half-plane and F_- is holomorphic on the lower half-plane, such that for x along the real axis,

$$F_+(x) - F_-(x) = f(x)$$

where $f(x)$ is some given real-valued function of $x \in \mathbb{R}$. The left-hand side of this equation may be understood either as the difference of the limits of $F_\pm$ from the appropriate half-planes, or as a hyperfunction distribution. Two functions of this form are a solution of the Riemann–Hilbert problem.

Formally, if $F_\pm$ solve the Riemann–Hilbert problem

$$f(x) = F_+(x) - F_-(x),$$

then the Hilbert transform of $f(x)$ is given by

$$H(f)(x) = \frac{1}{i}(F_+(x) + F_-(x))$$ (Pandey 1996, Chapter 2).

Hilbert Transform on the Circle

For a periodic function f the circular Hilbert transform is defined as

$$\tilde{f}(x) = \frac{1}{2\pi} \text{ p.v.}\int_0^{2\pi} f(t) \cot\left(\frac{x-t}{2}\right) dt.$$

The circular Hilbert transform is used in giving a characterization of Hardy space and in the study of the conjugate function in Fourier series. The kernel,

$$\cot\left(\frac{x-t}{2}\right)$$

is known as the Hilbert kernel since it was in this form the Hilbert transform was originally studied (Khvedelidze 2001).

The Hilbert kernel (for the circular Hilbert transform) can be obtained by making the Cauchy kernel $1/x$ periodic. More precisely, for $x \neq 0$

$$\frac{1}{2}\cot\left(\frac{x}{2}\right) = \frac{1}{x} + \sum_{n=1}^{\infty}\left(\frac{1}{x+2n\pi} + \frac{1}{x-2n\pi}\right).$$

Many results about the circular Hilbert transform may be derived from the corresponding results for the Hilbert transform from this correspondence.

Another more direct connection is provided by the Cayley transform $C(x) = (x - i) / (x + i)$, which carries the real line onto the circle and the upper half plane onto the unit disk. It induces a unitary map

$$Uf(x) = \pi^{-1/2}(x+i)^{-1} f(C(x))$$

of $L^2(\mathrm{T})$ onto $L^2(\mathrm{R})$. The operator U carries the Hardy space $H^2(\mathrm{T})$ onto the Hardy space $H^2(\mathrm{R})$.[2]

Hilbert Transform in Signal Processing

Bedrosian's Theorem: Bedrosian's theorem states that the Hilbert transform of the product of a low-pass and a high-pass signal with non-overlapping spectra is given by the product of the low-pass signal and the Hilbert transform of the high-pass signal, or

$$H(f_{LP}(t) f_{HP}(t)) = f_{LP}(t) H(f_{HP}(t))$$

where f_{LP} and f_{HP} are the low- and high-pass signals respectively (Schreier & Scharf 2010, 14).

Amplitude modulated signals are modelled as the product of a bandlimited "message" waveform, $u_m(t)$, and a sinusoidal "carrier":

$$u(t) = u_m(t) \cdot \cos(\omega t + \phi)$$

When $u_m(t)$ has no frequency content above the carrier frequency, $\frac{\omega}{2\pi}$ Hz, then by Bedrosian's theorem:

$$H(u)(t) = u_m(t) \cdot \sin(\omega t + \phi).$$ (Bedrosian 1962)

Causality

The function h with $h(t) = 1/(\pi t)$ is a non-causal filter and therefore cannot be implemented as is, if u is a time-dependent signal. If u is a function of a non-temporal variable, e.g., spatial, the non-causality might not be a problem. The filter is also of infinite support which may be a problem in certain applications. Another issue relates to what happens with the zero frequency (DC), which can be avoided by assuring that s does not contain a DC-component.

A practical implementation in many cases implies that a finite support filter, which in addition is made causal by means of a suitable delay, is used to approximate the computation. The approximation may also imply that only a specific frequency range is subject to the characteristic phase shift related to the Hilbert transform.

Discrete Hilbert Transform

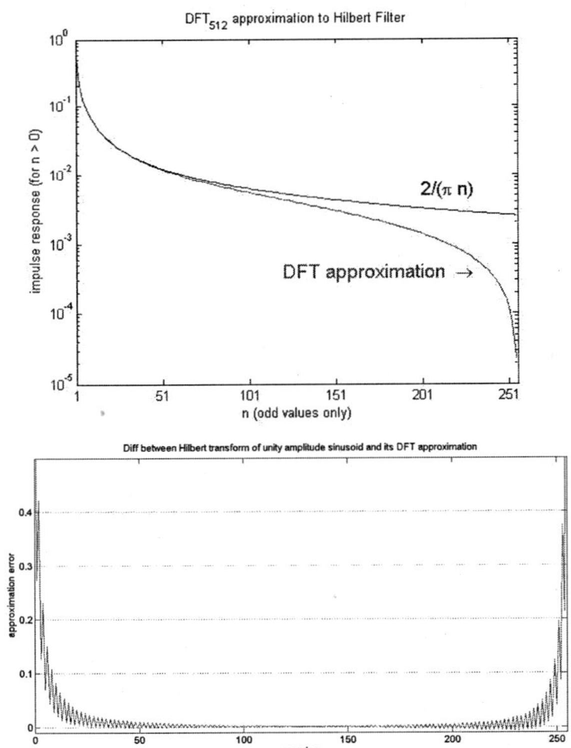

Figure: *The Hilbert transform of cos(wt) is sin(wt). This figure shows the difference between sin(wt) and an approximate Hilbert transform computed by the MATLAB library function, hilbert(---- ·)*

For a discrete function, u[n], with discrete-time Fourier transform (DTFT), U(ω), the Hilbert transform is given by:

$$H(u)[n] = DTFT^{-1}\{U(\omega)\cdot\sigma_H(\omega)\},$$

where:

$$\sigma_H(\omega) \stackrel{\text{def}}{=} \begin{cases} e^{+i\pi/2}, & -\pi < \omega < 0 \\ e^{-i\pi/2}, & 0 < \omega < \pi \\ 0, & \omega = -\pi, 0, \pi. \end{cases}$$

And by the convolution theorem, an equivalent formulation is:

$$H(u)[n] = u[n] * h[n],$$

where:

$$h[n] \stackrel{\text{def}}{=} DTFT^{-1}\{\sigma_H(\omega)\} = \begin{cases} 0, & \text{for } n \text{ even,} \\ \dfrac{2}{\pi n} & \text{for } n \text{ odd.} \end{cases}$$

When the convolution is performed numerically, an FIR approximation is substituted for $h[n]$, as shown in Figure, and we see rolloff of the passband at the low and high ends (0 and Nyquist), resulting in a bandpass filter. The high end can be restored, as shown in Figure, by an FIR that more closely resembles samples of the smooth, continuous-time $h(t)$. But as a practical matter, a properly-sampled $u[n]$ sequence has no useful components at those frequencies. As the impulse response gets longer, the low end frequencies are also restored.

With an FIR approximation to $h[n]$, a method called overlap-save is an efficient way to perform the convolution on a long $u[n]$ sequence. Sometimes the array FFT$\{h[n]\}$ is replaced by corresponding samples of $\sigma_H(\omega)$. That has the effect of convolving with the periodic summation:

$$h_N[n] \stackrel{\text{def}}{=} \sum_{m=-\infty}^{\infty} h[n - mN].$$

Figure compares a half-cycle of $h_N[n]$ with an equivalent length portion of $h[n]$. The difference between them and the fact that they are not shorter than the segment length (N) are sources of distortion that are managed (reduced) by increasing the segment length and overlap parameters.

The popular MATLAB function, hilbert(u,N), returns an approximate discrete Hilbert transform of $u[n]$ in the imaginary part of the complex output sequence. The real part is the original input sequence, so that the complex output is an analytic representation of $u[n]$. Similar to the discussion above, hilbert(u, N) only uses samples of the sgn(ω) distribution and therefore convolves with $h_N[n]$. Distortion can be managed by choosing N larger than the actual $u[n]$ sequence and discarding an appropriate number of output samples. An example of this type of distortion is shown in figure.

$$\frac{1}{n} * u[n] = \sum_{m=-\infty}^{\infty} \frac{u(m)}{n-m} \qquad m \neq n,$$

and showed that for $u(n)$ in ℓ^2 the sequence $H(u)[n]$ is also in ℓ^2. An elementary proof of this fact can be found in (Grafakos 1994). This transform was used by E. C. Titchmarsh to give alternate proofs of the results of M. Riesz in the continuous case, but it is not used for pragmatic signal processing.

Green's Function

In mathematics, a Green's function is a type of function used to solve inhomogeneous differential equations subject to specific initial conditions or boundary conditions. It is a kernel to the transform of a function's boundary-value to the function's value in some space (i.e., the response in the space). Under many-body theory, the term is also used in physics, specifically in quantum field theory, aerodynamics, aeroacoustics, electrodynamics and statistical field theory, to refer to various types of correlation functions, even those that do not fit the mathematical definition.

Green's functions are named after the British mathematician George Green, who first developed the concept in the 1830s. In the modern study of linear partial differential equations, Green's functions are studied largely from the point of view of fundamental solutions instead.

Definition and Uses

A Green's function, $G(x, s)$, of a linear differential operator $L = L(x)$ acting on distributions over a subset of the Euclidean space R^n, at a point s, is any solution of

$$LG(x,s) = \delta(x-s)$$

where δ is the Dirac delta function. This property of a Green's function can be exploited to solve differential equations of the form

$$Lu(x) = f(x).$$

If the kernel of L is non-trivial, then the Green's function is not unique. However, in practice, some combination of symmetry, boundary conditions and/or other externally imposed criteria will give a unique Green's function. Also, Green's functions in general are distributions, not necessarily proper functions.

Green's functions are also a useful tool in solving wave equations, diffusion equations, and in quantum mechanics, where the Green's function of the Hamiltonian is a key concept, with important links to the concept of density of states. As a side note, the Green's function as used in physics is usually defined with the opposite sign; that is,

$$LG(x,s) = -\delta(x-s).$$

This definition does not significantly change any of the properties of the Green's function.

If the operator is translation invariant, that is when L has constant coefficients with respect to x, then the Green's function can be taken to be a convolution operator, that is,

$$G(x,s) = G(x-s).$$

In this case, the Green's function is the same as the impulse response of linear time-invariant system theory.

Motivation

Loosely speaking, if such a function G can be found for the operator L, then if we multiply the equation (1) for the Green's function by $f(s)$, and then perform an integration in the s variable, we obtain;

$$\int LG(x,s)f(s)ds = \int \delta(x-s)f(s)ds = f(x).$$

The right hand side is now given by the equation (2) to be equal to $L\ u(x)$, thus:

$$Lu(x) = \int LG(x,s)f(s)ds.$$

Because the operator $L = L(x)$ is linear and acts on the variable x alone (not on the variable of integration s), we can take the operator L outside of the integration on the right hand side, obtaining;

$$Lu(x) = L\left(\int G(x,s)f(s)ds\right).$$

And this suggests;

$$u(x) = \int G(x,s)f(s)ds.$$

Thus, we can obtain the function $u(x)$ through knowledge of the Green's function in equation (1), and the source term on the right hand side in equation (2). This process relies upon the linearity of the operator L.

In other words, the solution of equation (2), $u(x)$, can be determined by the integration given in equation (3). Although $f(x)$ is known, this integration cannot be performed unless G is also known. The problem now lies in finding the Green's function G that satisfies equation (1). For this reason, the Green's function is also sometimes called the fundamental solution associated to the operator L.

Not every operator L admits a Green's function. A Green's function can also be thought of as a right inverse of L. Aside from the difficulties of finding a Green's function for a particular operator, the integral in equation (3), may be quite difficult to evaluate. However the method gives a theoretically exact result.

This can be thought of as an expansion of f according to a Dirac delta function basis (projecting f over $\delta(x$ " $s)$) and a superposition of the solution on each projection. Such an integral equation is known as a Fredholm integral equation, the study of which constitutes Fredholm theory.

Green's Functions for Solving Inhomogeneous Boundary Value Problems

The primary use of Green's functions in mathematics is to solve non-homogeneous boundary value problems. In modern theoretical physics, Green's functions are also usually used as propagators in Feynman diagrams (and the phrase *Green's function* is often used for any correlation function).

Framework

Let L be the Sturm–Liouville operator, a linear differential operator of the form

$$L = \frac{d}{dx}\left[p(x)\frac{d}{dx}\right] + q(x)$$

and let D be the boundary conditions operator

$$Du = \begin{cases} \alpha_1 u'(0) + \beta_1 u(0) \\ \alpha_2 u'(l) + \beta_2 u(l). \end{cases}$$

Let $f(x)$ be a continuous function in [0,l]. We shall also suppose that the problem

$$\begin{aligned} Lu &= f \\ Du &= 0 \end{aligned}$$

is regular (i.e., only the trivial solution exists for the homogeneous problem).

Theorem

There is one and only one solution $u(x)$ which satisfies

$$\begin{aligned} Lu &= f \\ Du &= 0 \end{aligned}$$

and it is given by

$$u(x) = \int_0^{\ell} f(s)G(x,s)ds$$

where $G(x,s)$ is a Green's function satisfying the following conditions:

1. $G(x,s)$ is continuous in x and s
2. For $x \neq s$, $LG(x,s) = 0$
3. For $s \neq 0$, $DG(x,s) = 0$
4. Derivative "jump": $G'(s_{+0},s) - G'(s_{-0},s) = 1/p(s)$
5. Symmetry: $G(x, s) = G(s, x)$.

Finding Green's Functions

Eigenvalue Expansions: If a differential operator L admits a set of eigenvectors $\Psi_n(x)$ (i.e., a set of functions Ψ_n and scalars λ_n such that $L\Psi_n = \lambda_n \Psi_n$) that is complete, then it is possible to construct a Green's function from these eigenvectors and eigenvalues.

Complete means that the set of functions $\{\Psi_n\}$ satisfies the following completeness relation:

$$\delta(x - x') = \sum_{n=0}^{\infty} \Psi_n^{\dagger}(x)\Psi_n(x').$$

Then the following holds:

$$G(x,x') = \sum_{n=0}^{\infty} \frac{\Psi_n^{\dagger}(x)\Psi_n(x')}{\lambda_n}$$

where $\dagger$ represents complex conjugation.

Applying the operator L to each side of this equation results in the completeness relation, which was assumed true.

The general study of the Green's function written in the above form, and its relationship to the function spaces formed by the eigenvectors, is known as Fredholm theory.

There are several other methods for finding Green's functions, including the method of images, separation of variables, and Laplace transforms (Cole 2011).

Chapter 4

Bounded Variation

In mathematical analysis, a function of bounded variation, also known as a BV function, is a real-valued function whose total variation is bounded (finite): the graph of a function having this property is well behaved in a precise sense. For a continuous function of a single variable, being of bounded variation means that the distance along the direction of the y-axis, neglecting the contribution of motion along x-axis, travelled by a point moving along the graph has a finite value. For a continuous function of several variables, the meaning of the definition is the same, except for the fact that the continuous path to be considered cannot be the whole graph of the given function (which is a hypersurface in this case), but can be every intersection of the graph itself with a hyperplane (in the case of functions of two variables, a plane) parallel to a fixed x-axis and to the y-axis.

Functions of bounded variation are precisely those with respect to which one may find Riemann–Stieltjes integrals of all continuous functions.

Another characterization states that the functions of bounded variation on a closed interval are exactly those f which can be written as a difference $g - h$, where both g and h are bounded monotone.

In the case of several variables, a function f defined on an open subset Ω of $\mathbb{R}^n$ is said to have bounded variation if its distributional derivative is a finite vector Radon measure.

One of the most important aspects of functions of bounded variation is that they form an algebra of discontinuous functions whose first derivative exists almost everywhere: due to this fact, they can and frequently are used to define generalized solutions of nonlinear problems involving functionals, ordinary and partial differential equations in mathematics, physics and engineering. Considering the problem of multiplication of distributions or more generally the problem of defining general nonlinear operations on generalized functions, *function of bounded variation are the smallest algebra which has to be embedded in every space of generalized functions preserving the result of multiplication.*

History

According to Boris Golubov, *BV* functions of a single variable were first introduced by Camille Jordan, in the paper (Jordan 1881) dealing with the convergence of Fourier series. The first successful step in the generalization of this concept to functions of several variables was due to Leonida Tonelli, who introduced a class of *continuous BV* functions in 1926 (Cesari 1986, pp. 47–48), to extend his direct method for finding solutions to problems in the calculus of variations in more than one variable.

Ten years after, in (Cesari 1936), Lamberto Cesari *changed the continuity requirement* in Tonelli's definition *to a less restrictive integrability requirement*, obtaining for the first time the class of functions of bounded variation of several variables in its full generality: as Jordan did before him, he applied the concept to resolve of a problem concerning the convergence of Fourier series, but for functions of *two variables*. After him, several authors applied *BV* functions to study Fourier series in several variables, geometric measure theory, calculus of variations, and mathematical physics.

Renato Caccioppoli and Ennio de Giorgi used them to define measure of nonsmooth boundaries of sets. Olga Arsenievna Oleinik introduced her view of generalized solutions for nonlinear partial differential equations as functions from the space *BV* in the paper (Oleinik 1957), and was able to construct a generalized solution of bounded variation of a first order partial differential equation in the paper (Oleinik 1959): few years later, Edward D. Conway and Joel A. Smoller applied *BV*-functions to the study of a single nonlinear hyperbolic partial differential equation of first order in the paper (Conway &

Smoller 1966), proving that the solution of the Cauchy problem for such equations is a function of bounded variation, provided the initial value belongs to the same class. Aizik Isaakovich Vol'pert developed extensively a calculus for *BV* functions: in the paper (Vol'pert 1967) he proved the chain rule for BV functions and in the book (Hudjaev & Vol'pert 1985) he, jointly with his pupil Sergei Ivanovich Hudjaev, explored extensively the properties of *BV* functions and their application. His chain rule formula was later extended by Luigi Ambrosio and Gianni Dal Maso in the paper (Ambrosio & Dal Maso 1990).

Formal Definition

BV Functions of one Variable: *Definition 1.1.* The total variation of a real-valued (or more generally complex-valued) function f, defined on an interval $[a, b]\subset\mathbb{R}$ is the quantity

$$V_b^a(f) = \sup_{P\in\mathcal{P}} \sum_{i=0}^{n_P-1} |f(x_{i+1}) - f(x_i)|.$$

where the supremum is taken over the set $\mathcal{P} = \left\{P = \{x_0,\ldots,x_{n_P}\} \mid P \text{ is a partition of } [a,b]\right\}$ of all partitions of the interval considered.

If f is differentiable and its derivative is Riemann-integrable, its total variation is the vertical component of the arc-length of its graph, that is to say,

$$V_b^a(f) = \int_a^b |f'(x)|\mathrm{d}x.$$

Definition 1.2. A real-valued function f on the real line is said to be of bounded variation (BV function) on a chosen interval $[a, b]\subset\mathbb{R}$ if its total variation is finite, *i.e.*

$$f \in BV([a,b]) \Leftrightarrow V_b^a(f) < +\infty$$

It can be proved that a real function f is of bounded variation in an interval if and only if it can be written as the difference $f = f_1 - f_2$ of two non-decreasing functions: this result is known as the Jordan decomposition.

Through the Stieltjes integral, any function of bounded variation on a closed interval $[a, b]$ defines a bounded linear functional on $C([a, b])$. In this special case, the Riesz representation theorem states that every bounded linear functional arises uniquely in this way. The normalised positive functionals or probability measures correspond to positive non-decreasing lower semicontinuous functions. This point of view has been important in spectral theory, in particular in its application to ordinary differential equations.

BV *Functions of Several Variables*

Functions of bounded variation, BV functions, are functions whose distributional derivative is a finite Radon measure. More precisely:

Definition 2.1. Let Ω be an open subset of $\mathbb{R}^n$. A function u belonging to $L^1(\Omega)$ is said of bounded variation (BV function), and write

$$u \in BV(\Omega)$$

if there exists a finite vector Radon measure $Du \in \mathcal{M}(\Omega, \mathbb{R}^n)$ such that the following equality holds

$$\int_\Omega u(x)\operatorname{div}\phi(x)\mathrm{d}x = -\int_\Omega \langle \phi, Du(x) \rangle \qquad \forall \phi \in C_c^1(\Omega, \mathbb{R}^n)$$

that is, $u$defines a linear functional on the space $C_c^1(\Omega, \mathbb{R}^n)$ of continuously differentiable vector functions $\phi$of compact support contained in Ω: the vector measure Du represents therefore the distributional or weak gradient of u.

An equivalent definition is the following.

Definition: Given a function u belonging to $L^1(\Omega)$, the total variation of in is defined as

$$V(u,\Omega) := \sup\left\{\int_\Omega u(x)\operatorname{div}\phi(x)\mathrm{d}x : \phi \in C_c^1(\Omega, \mathbb{R}^n), \|\phi\|_{L^\infty(\Omega)} \leq 1\right\}$$

where $\| \; \|_{L^\infty(\Omega)}$ is the essential supremum norm. Sometimes, especially in the theory of Caccioppoli sets, the following notation is used

$$\int_\Omega |Du| = V(u,\Omega)$$

in order to emphasize that $V(u,\Omega)$ is the total variation of the distributional / weak gradient of u. This notation reminds also that if $u$is of class C^1 (i.e. a continuous and differentiable function having continuous derivatives) then its variation is exactly the integral of the absolute value of its gradient.

The space of functions of bounded variation (BV functions) can then be defined as

$$BV(\Omega) = \{u \in L^1(\Omega) : V(u,\Omega) < +\infty\}$$

The two definitions are equivalent since if $V(u,\Omega) < +\infty$ then

$$\left|\int_\Omega u(x)\mathrm{div}\phi(x)\mathrm{d}x\right| \le V(u,\Omega) \, \|\phi\|_{L^\infty(\Omega)} \quad \forall\phi \in C_c^1(\Omega,\mathbb{R}^n)$$

therefore $\int_\Omega u(x)\mathrm{div}\phi(x)$ defines a continuous linear functional on the space $C_c^1(\Omega,\mathbb{R}^n)$. Since $C_c^1(\Omega,\mathbb{R}^n) \subset C^0(\Omega,\mathbb{R}^n)$ as a linear subspace, this continuous linear functional can be extended continuously and linearily to the whole $C^0(\Omega,\mathbb{R}^n)$ by the Hahn–Banach theorem i.e. it defines a Radon measure.

Locally BV *Functions*

If the function space of locally integrable functions, i.e. functions belonging to $L^1_{loc}(\Omega)$, is considered in the preceding definitions 1.2, 2.1 and 2.2 instead of the one of globally integrable functions, then the function space defined is that of functions of locally bounded variation. Precisely, developing this idea for definition 2.2, a local variation is defined as follows,

$$V(u,U) := \sup\left\{\int_\Omega u(x)\mathrm{div}\phi(x)\mathrm{d}x : \phi \in C_c^1(U,\mathbb{R}^n), \|\phi\|_{L^\infty(\Omega)} \le 1\right\}$$

for every set $U \in \mathcal{O}_c(\Omega)$, having defined $\mathcal{O}_c(\Omega)$ as the set of all precompact open subsets of Ω with respect to the standard topology of finite dimensional vector spaces, and correspondingly the class of functions of locally bounded variation is defined as

$$BV_{loc}(\Omega) = \{u \in L^1_{loc}(\Omega) : V(u,U) < +\infty \ \forall U \in \mathcal{O}_c(\Omega)\}$$

Notation

There are basically two distinct conventions for the notation of spaces of functions of locally or globally bounded variation, and unfortunately they are quite similar: the first one, which is the one adopted in this entry, is used for example in references Giusti (1984) (partially), Hudjaev & Vol'pert (1985) (partially), Giaquinta, Modica & Souèek (1998) and is the following one

- $BV(\Omega)$ identifies the space of functions of globally bounded variation
- $BV_{loc}(\Omega)$ identifies the space of functions of locally bounded variation

The second one, which is adopted in references Vol'pert (1967) and Maz'ya (1985) (partially), is the following:

- $\overline{BV}(\Omega)$ identifies the space of functions of globally bounded variation
- $BV(\Omega)$ identifies the space of functions of locally bounded variation

Basic Properties

Only the properties common to functions of one variable and to functions of several variables will be considered in the following, and proofs will be carried on only for functions of several variables since the proof for the case of one variable is a straightforward adaptation of the several variables case: also, in each section it will be stated if the property is shared also by functions of locally bounded variation or not. References (Giusti 1984, pp. 7–9), (Hudjaev & Vol'pert 1985) and (Màlek et al. 1996) are extensively used.

Examples

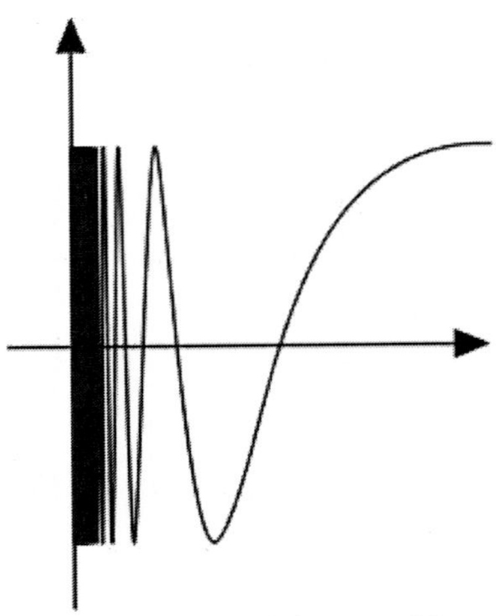

Figure: *The function* f(x)=*sin(1/x) is* not *of bounded variation on the interval* $[0,2/\pi]$.

The function

$$f(x)=\begin{cases}0, & \text{if } x=0\\ \sin(1/x), & \text{if } x\neq 0\end{cases}$$

is *not* of bounded variation on the interval $[0,2/\pi]$

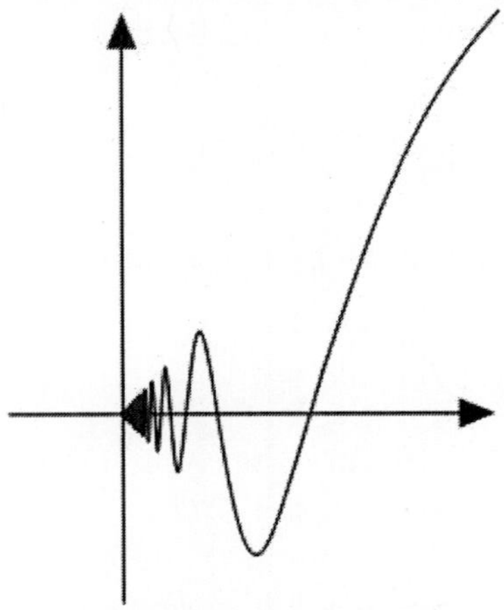

Figure: *The function* f(x)=x *sin(1/*x*) is* not *of bounded variation on the interval* $[0,2/\pi]$.

While it is harder to see, the continuous function

$$f(x)=\begin{cases}0, & \text{if } x=0\\ x\sin(1/x), & \text{if } x\neq 0\end{cases}$$

is *not* of bounded variation on the interval $[0,2/\pi]$ either.

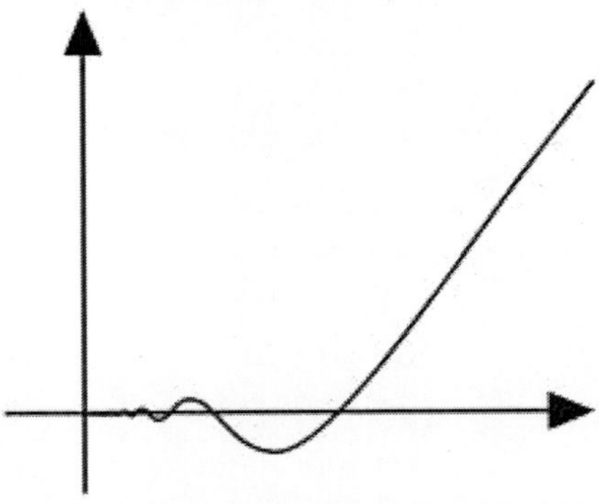

Figure: *The function* f(x)=x^2 *sin(1/*x*)* is *of bounded variation on the interval* $[0,2/\pi]$.

At the same time, the function

$$f(x) = \begin{cases} 0, & \text{if } x = 0 \\ x^2 \sin(1/x), & \text{if } x \neq 0 \end{cases}$$

is of bounded variation on the interval $[0, 2/\pi]$. However, *all three functions are of bounded variation on each interval* $[a,b]$ *with* $a > 0$.

The Sobolev space $W^{1,1}(\Omega)$ is a proper subset of $BV(\Omega)$. In fact, for each u in $W^{1,1}(\Omega)$ it is possible to choose a measure $\mu := \nabla u \mathcal{L}$ (where $\mathcal{L}$ is the Lebesgue measure on Ω) such that the equality

$$\int u \operatorname{div}\phi = -\int \phi \, d\mu = -\int \phi \nabla u \qquad \forall \phi \in C_c^1$$

holds, since it is nothing more than the definition of weak derivative, and hence holds true. One can easily find an example of a BV function which is not $W^{1,1}$: in dimension one, any step function with a non-trivial jump will do.

Applications

Mathematics: Functions of bounded variation have been studied in connection with the set of discontinuities of functions and differentiability of real functions, and the following results are well-known. If f is a real function of bounded variation on an interval $[a,b]$ then

- f is continuous except at most on a countable set;
- f has one-sided limits everywhere (limits from the left everywhere in $(a,b]$, and from the right everywhere in $[a,b)$;
- the derivative $f'(x)$ exists almost everywhere (i.e. except for a set of measure zero).

For real functions of several real variables

- the characteristic function of a Caccioppoli set is a BV function: BV functions lie at the basis of the modern theory of perimeters.
- Minimal surfaces are graphs of BV functions.

Physics and Engineering

The ability of *BV* functions to deal with discontinuities has made their use widespread in the applied sciences: solutions of problems in mechanics, physics, chemical kinetics are very often representable by functions of bounded variation. The book (Hudjaev & Vol'pert 1985) details a very ample set of mathematical physics applications of *BV* functions. Also there is some modern application which deserves a brief description.

- The Mumford-Shah Functional: the segmentation problem for a two-dimensional image, i.e. the problem of faithful reproduction of contours and grey scales is equivalent to the minimization of such functional.

Second Derivative

In calculus, the second derivative of a function f is the derivative of the derivative of f. Roughly speaking, the second derivative measures how the rate of change of a quantity is itself changing; for example, the second derivative of the position of a vehicle with respect to time is the instantaneous acceleration of the vehicle, or the rate at which the velocity of the vehicle is changing.

On the graph of a function, the second derivative corresponds to the curvature or concavity of the graph. The graph of a function with positive second derivative curves upwards, while the graph of a function with negative second derivative curves downwards.

Second Derivative Power Rule

The power rule for the first derivative, if solved down a bit, will produce the second derivative power rule. The rule is given below:

$$\frac{d^2}{dx^2}[x^n] = n(n-1)x^{(n-2)} = (n^2-n)x^{(n-2)}.$$

Notation

The second derivative of a function $f(x)$ is usually denoted $f''(x)$. That is:

$$f'' = (f')'$$

When using Leibniz's notation for derivatives, the second derivative of a dependent variable y with respect to an independent variable x is written

$$\frac{d^2y}{dx^2}.$$

This notation is derived from the following formula:

$$\frac{d^2y}{dx^2}=\frac{d}{dx}\left(\frac{dy}{dx}\right).$$

Example

Given the function

$$f(x)=x^3,$$

the derivative of f is the function

$$f'(x)=3x^2.$$

The second derivative of f is the derivative of f2, namely

$$f''(x)=6x.$$

Relation to the Graph

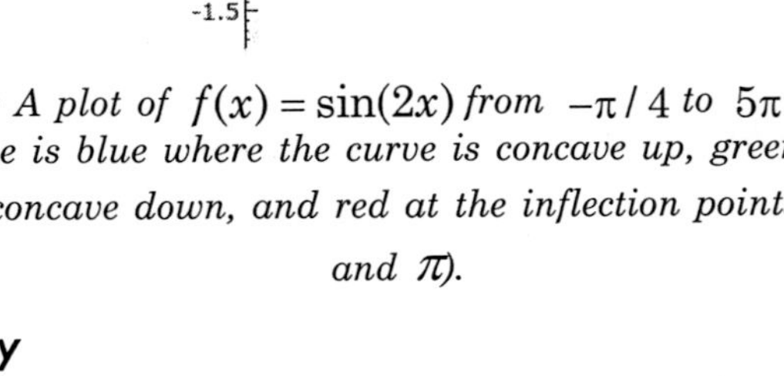

Figure: *A plot of* $f(x)=\sin(2x)$ *from* $-\pi/4$ *to* $5\pi/4$. *The tangent line is blue where the curve is concave up, green where the curve is concave down, and red at the inflection points (0, $\pi/2$, and π).*

Concavity

The second derivative of a function f measures the concavity of the graph of f. A function whose second derivative is positive will be concave up (sometimes referred to as convex), meaning that the tangent line will lie below the graph of the function. Similarly, a function whose second derivative is negative will be concave down (sometimes called simply "concave"), and its tangent lines will lie above the graph of the function.

Inflection Point

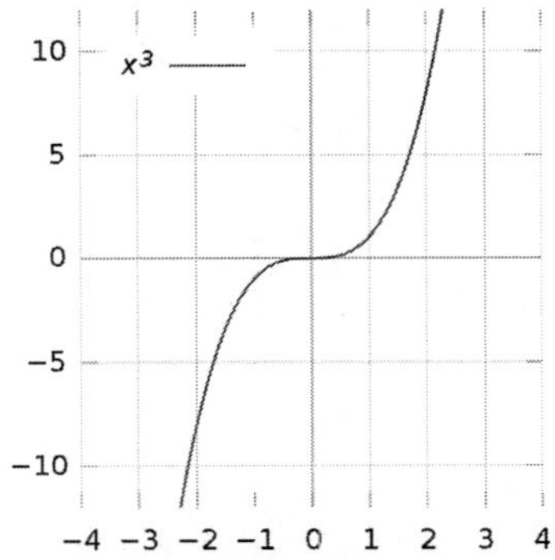

Figure: *Plot of* $y = x^3$ *with inflection point of (0,0), also a saddle point.*

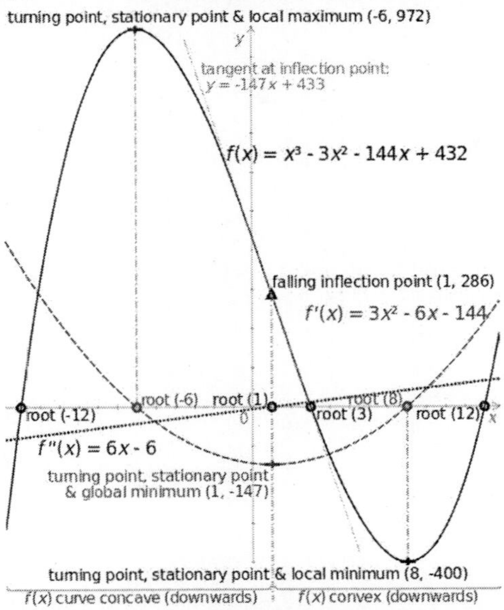

Figure: *Graph showing the relationship between the roots, turning points, stationary points, inflection point and concavity of a cubic polynomial* $x^3 - 3x^2 - 144x + 432$ *and its first and second derivatives.*

In differential calculus, an inflection point, point of inflection, flex, or inflection (inflexion) is a point on a curve at which the curvature or concavity changes sign from plus to minus or from minus to plus. The curve changes from being concave upwards (positive curvature) to concave downwards (negative curvature), or vice versa. If one imagines driving a vehicle along a winding road, inflection is the point at which the steering-wheel is momentarily "straight" when being turned from left to right or vice versa. A point where the curvature vanishes but does not change sign is sometimes called a point of undulation or undulation point.

In algebraic geometry an inflection point is defined slightly more generally, as a point where the tangent meets the curve to order at least 3, and an undulation point or hyperflex is defined as a point where the tangent meets the curve to order at least 4.

Equivalent Forms

The following are all equivalent to the above definition:

- a point on a curve at which the second derivative changes sign. This is very similar to the previous definition, since the sign of the curvature is always the same as the sign of the second derivative, but note that the curvature is not the same as the second derivative.
- a point (x, y) on a function, $f(x)$, at which the first derivative, $f2\ (x)$, is at an extremum, i.e. a (local) minimum or maximum. (This is not the same as saying that y is at an extremum).
- a point p on a curve at which the tangent crosses the curve at that point. For an algebraic curve, this means a non singular point where the multiplicity of the intersection at p of the tangent line and the curve is odd and greater than 2.

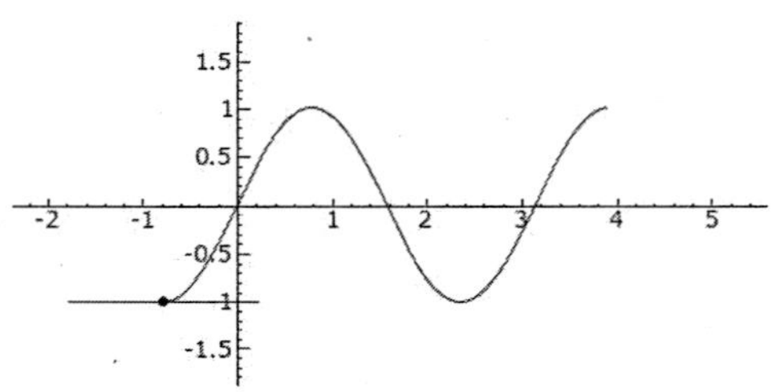

Figure: Plot of $f(x) = \sin(2x)$ from $-\pi/4$ to $5\pi/4$; note f's second derivative is $f3\ (x) = -4\sin(2x)$. Tangent is blue where the curve is concave up (above its own tangent), green where concave down (below its tangent), and red at inflection points: 0, $\pi/2$ and π

A Necessary but not Sufficient Condition

If x is an inflection point for f then the second derivative, $f3$ (x), is equal to zero if it exists, but this condition does not provide a sufficient definition of a point of inflection. One also needs the lowest-order (above the second) non-zero derivative to be of odd order (third, fifth, etc.). If the lowest-order non-zero derivative is of even order, the point is not a point of inflection, but an *undulation point*. However, in algebraic geometry, both inflection points and undulation points are usually called *inflection points*. An example of such an undulation point is $y = x^4$ for x=0.

It follows from the definition that the sign of $f2$ (x) on either side of the point (x,y) must be the same. If this is positive, the point is a *rising point of inflection*; if it is negative, the point is a *falling point of inflection.*

Categorization of Points of Inflection

Points of inflection can also be categorised according to whether $f2$ (x) is zero or not zero.

- if $f2$ (x) is zero, the point is a *stationary point of inflection*, also known as a saddle-point
- if $f2$ (x) is not zero, the point is a *non-stationary point of inflection*

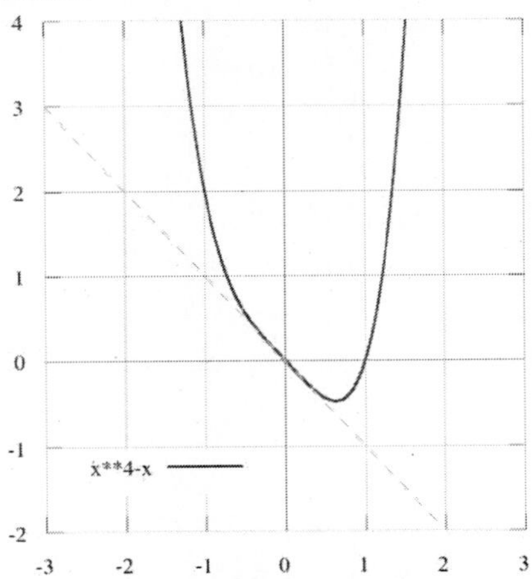

Figure: *$y = x^4 - x$ has a 2nd derivative of zero at point (0,0), but it is not an inflection point because the fourth derivative is the first higher order non-zero derivative (the third derivative is zero as well).*

An example of a saddle point is the point (0,0) on the graph $y = x^3$. The tangent is the x-axis, which cuts the graph at this point.

A non-stationary point of inflection can be visualised if the graph $y = x^3$ is rotated slightly about the origin. The tangent at the origin still cuts the graph in two, but its gradient is non-zero.

Note that an inflection point is also called an *ogee*, although this term is sometimes applied to the entire curve which contains an inflection point.

Asymptotic Functions

Some functions change concavity without having points of inflection. Instead, they can change concavity around vertical asymptotes or discontinuities. Take, for example, the function $2x^2/(x^2 - 1)$. It is concave up when $|x| > 1$ and concave down when $|x| < 1$. However, it has no points of inflection because 1 and -1 are not in the domain of the function.

Second Derivative Test

In calculus, the second derivative test is a criterion for determining whether a given critical point of a real function of one variable is a local maximum or a local minimum using the value of the second derivative at the point.

The test states: if the function f is twice differentiable at a critical point x (i.e. $f'(x) = 0$), then:

- If $f''(x) < 0$ then f has a local maximum at x.
- If $f''(x) > 0$ then f has a local minimum at x.
- If $f''(x) = 0$, the test is inconclusive.

In the latter case, Taylor's Theorem may be used to determine the behaviour of f near x using higher derivatives.

Hessian Matrix

In mathematics, the Hessian matrix (or simply the Hessian) is the square matrix of second-order partial derivatives of a function; that is, it describes the local curvature of a function of many variables. The Hessian matrix was developed in the 19th century by the German mathematician Ludwig Otto Hesse and later named after him. Hesse himself had used the term "functional determinants".

Given the real-valued function

$$f(x_1, x_2, \ldots, x_n),$$

if all second partial derivatives of f exist and are continuous over the domain of the function, then the Hessian matrix of f is the matrix

$$H(f)_{ij}(\mathrm{x}) = D_i D_j f(\mathrm{x})$$

where $x = (x_1, x_2, \ldots, x_n)$ and D_i is the differentiation operator with respect to the ith argument and the Hessian becomes

$$H(f) = \begin{bmatrix} \frac{\partial^2 f}{\partial x_1^2} & \frac{\partial^2 f}{\partial x_1\,\partial x_2} & \cdots & \frac{\partial^2 f}{\partial x_1\,\partial x_n} \\ [2.2ex]\frac{\partial^2 f}{\partial x_2\,\partial x_1} & \frac{\partial^2 f}{\partial x_2^2} & \cdots & \frac{\partial^2 f}{\partial x_2\,\partial x_n} \\ [2.2ex]\vdots & \vdots & \ddots & \vdots \\ [2.2ex]\frac{\partial^2 f}{\partial x_n\,\partial x_1} & \frac{\partial^2 f}{\partial x_n\,\partial x_2} & \cdots & \frac{\partial^2 f}{\partial x_n^2} \end{bmatrix}.$$

Because f is often clear from context, $H(f)(\mathrm{x})$ is frequently shortened to simply $H(\mathrm{x})$.

The Hessian matrix is related to the Jacobian matrix by, $H(f)(\mathrm{x}) = J(\nabla f)(\mathrm{x})$.

The determinant of the above matrix is also sometimes referred to as the Hessian.[1]

Hessian matrices are used in large-scale optimization problems within Newton-type methods because they are the coefficient of the quadratic term of a local Taylor expansion of a function. That is,

$$y = f(\mathrm{x} + \Delta\mathrm{x}) \approx f(\mathrm{x}) + J(\mathrm{x})\Delta\mathrm{x} + \frac{1}{2}\Delta\mathrm{x}^{\mathrm{T}} H(\mathrm{x})\Delta\mathrm{x}$$

where J is the Jacobian matrix, which is a vector (the gradient) for scalar-valued functions. The full Hessian matrix can be difficult to compute in practice; in such situations, quasi-Newton algorithms have been developed that use approximations to the Hessian. The best-known quasi-Newton algorithm is the BFGS algorithm.

Mixed Derivatives and Symmetry of the Hessian

The mixed derivatives of f are the entries *off* the main diagonal in the Hessian. Assuming that they are continuous, the order of differentiation does not matter (Clairaut's theorem). For example,

$$\frac{\partial}{\partial x}\left(\frac{\partial f}{\partial y}\right)=\frac{\partial}{\partial y}\left(\frac{\partial f}{\partial x}\right).$$

This can also be written as:

$$f_{yx}=f_{xy}.$$

In a formal statement: if the second derivatives of f are all continuous in a neighbourhood D, then the Hessian of f is a symmetric matrix throughout D.

Critical Points and Discriminant

If the gradient of f (i.e. its derivative in the vector sense) is zero at some point x, then f has a *critical point* (or *stationary point*) at x. The determinant of the Hessian at x is then called the discriminant. If this determinant is zero then x is called a *degenerate critical point* of f, this is also called a *non-Morse critical point* of f. Otherwise it is non-degenerate, this is called a *Morse critical point* of f.

The following test can be applied at a non-degenerate critical point x. If the Hessian is positive definite at x, then f attains a local minimum at x. If the Hessian is negative definite at x, then f attains a local maximum at x. If the Hessian has both positive and negative eigenvalues then x is a saddle point for f (this is true even if x is degenerate). Otherwise the test is inconclusive.

Note that for positive semidefinite and negative semidefinite Hessians the test is inconclusive (yet a conclusion can be made that f is locally convex or concave respectively). However, more can be said from the point of view of Morse theory.

In view of what has just been said, the second derivative test for functions of one and two variables is simple. In one variable, the Hessian contains just one second derivative; if it is positive then x is a local minimum, if it is negative then x is a local maximum; if it is zero then the test is inconclusive. In two variables, the determinant can be used, because the determinant is the product of the eigenvalues. If it is positive then the eigenvalues are both positive, or both negative. If it is negative then the two eigenvalues have different signs. If it is zero, then the second derivative test is inconclusive.

More generally, the second-order conditions that are sufficient for a local minimum or maximum can be expressed in terms of the sequence of principal (upper-leftmost) minors (determinants of sub-matrices) of the Hessian; these conditions are a special case of those given in the next section for bordered Hessians for constrained optimization—the case in which the number of constraints is zero.

Bordered Hessian

A bordered Hessian is used for the second-derivative test in certain constrained optimization problems. Given the function as before:

$$f(x_1, x_2, \ldots, x_n),$$

but adding a constraint function such that:

$$g(x_1, x_2, \ldots, x_n) = c,$$

the bordered Hessian appears as

$$H(f,g) = \begin{bmatrix} 0 & \dfrac{\partial g}{\partial x_1} & \dfrac{\partial g}{\partial x_2} & \cdots & \dfrac{\partial g}{\partial x_n} \\ \text{[2.2}ex\text{]}\dfrac{\partial g}{\partial x_1} & \dfrac{\partial^2 f}{\partial x_1^2} & \dfrac{\partial^2 f}{\partial x_1 \partial x_2} & \cdots & \dfrac{\partial^2 f}{\partial x_1 \partial x_n} \\ \text{[2.2}ex\text{]}\dfrac{\partial g}{\partial x_2} & \dfrac{\partial^2 f}{\partial x_2 \partial x_1} & \dfrac{\partial^2 f}{\partial x_2^2} & \cdots & \dfrac{\partial^2 f}{\partial x_2 \partial x_n} \\ \text{[2.2}ex\text{]}\vdots & \vdots & \vdots & \ddots & \vdots \\ \text{[2.2}ex\text{]}\dfrac{\partial g}{\partial x_n} & \dfrac{\partial^2 f}{\partial x_n \partial x_1} & \dfrac{\partial^2 f}{\partial x_n \partial x_2} & \cdots & \dfrac{\partial^2 f}{\partial x_n^2} \end{bmatrix}$$

If there are, say, m constraints then the zero in the north-west corner is an $m \times m$ block of zeroes, and there are m border rows at the top and m border columns at the left.

The above rules stating that extrema are characterized by a positive definite or negative definite Hessian cannot apply here since a bordered Hessian cannot be definite: we have

$z'Hz = 0$ if vector z has a non-zero as its first element, followed by zeroes.

The second derivative test consists here of sign restrictions of the determinants of a certain set of $n - m$ submatrices of the bordered Hessian. Intuitively, one can think of the m constraints as reducing the problem to one with $n - m$ free variables. (For example, the maximization of $f(x_1,x_2,x_3)$ subject to the constraint $x_1 + x_2 + x_3 = 1$ can be reduced to the maximization of $f(x_1,x_2,1-x_1-x_2)$ without constraint.)

Specifically, sign conditions are imposed on the sequence of principal minors (determinants of upper-left-justified sub-matrices) of the bordered Hessian, the smallest minor consisting of the truncated first m+1 rows and columns, the next consisting of the truncated first m+2 rows and columns, and so on, with the last being the entire bordered Hessian. There are thus n–m minors to consider. A sufficient condition for a local *maximum* is that these minors alternate in sign with the largest one having the sign of $(-1)^{m+1}$. A sufficient condition for a local *minimum* is that all of these minors have the sign of $(-1)^m$. (Note that in the unconstrained case of m=0 these conditions coincide with the conditions for the unbordered Hessian to be negative definite or positive definite respectively.)

Vector-valued Functions

If f is instead a function from $\mathbb{R}^n \to \mathbb{R}^m$, i.e.

$$f(x_1,x_2,\ldots,x_n) = (f_1,f_2,\ldots,f_m),$$

then the array of second partial derivatives is not a two-dimensional matrix of size $n \times n$, but rather a tensor of order 3. This can be thought of as a multi-dimensional array with dimensions $m \times n \times n$, which degenerates to the usual Hessian matrix for $m = 1$.

Laplace Operator

In mathematics the Laplace operator or Laplacian is a differential operator given by the divergence of the gradient of a function on Euclidean space. It is usually denoted by the symbols $\nabla \cdot \nabla$, ∇^2 or ∇. The Laplacian $\nabla f(p)$ of a function f at a point p, up to a constant depending on the dimension, is the rate at which the average value of f over spheres centred at p, deviates from $f(p)$ as the radius of the sphere grows. In a Cartesian coordinate system, the Laplacian is given by sum of second partial derivatives of the function with respect to each independent variable. In other coordinate systems such as cylindrical and spherical coordinates, the Laplacian also has a useful form.

The Laplace operator is named after the French mathematician Pierre-Simon de Laplace (1749–1827), who first applied the operator to the study of celestial mechanics, where the operator gives a constant multiple of the mass density when it is applied to a given gravitational potential. Solutions of the equation $\nabla f = 0$, now called Laplace's equation, are the so-called harmonic functions, and represent the possible gravitational fields in free space.

The Laplacian occurs in differential equations that describe many physical phenomena, such as electric and gravitational potentials, the diffusion equation for heat and fluid flow, wave propagation, and quantum mechanics. The Laplacian represents the flux density of the gradient flow of a function. For instance, the net rate at which a chemical dissolved in a fluid moves towards or away from some point is proportional to the Laplacian of the chemical concentration at that point; expressed symbolically, the resulting equation is the diffusion equation. For these reasons, it is extensively used in the sciences for modelling all kinds of physical phenomena. The Laplacian is the simplest elliptic operator, and is at the core of Hodge theory as well as the results of de Rham cohomology. In image processing and computer vision, the Laplacian operator has been used for various tasks such as blob and edge detection.

Definition

The Laplace operator is a second order differential operator in the n-dimensional Euclidean space, defined as the divergence ($\nabla \cdot$) of the gradient (∇f). Thus if f is a twice-differentiable real-valued function, then the Laplacian of f is defined by

$$\Delta f = \nabla^2 f = \nabla \cdot \nabla f$$

where the latter notations derive from formally writing $\nabla = \left(\frac{\partial}{\partial x_1}, \ldots, \frac{\partial}{\partial x_n}\right)$. Equivalently, the Laplacian of f is the sum of all the *unmixed* second partial derivatives in the Cartesian coordinates x_i :

$$\Delta f = \sum_{i=1}^{n} \frac{\partial^2 f}{\partial x_i^2}$$

As a second-order differential operator, the Laplace operator maps C^k-functions to C^{k-2}-functions for $k \geq 2$. The expression (1) (or equivalently (2)) defines an operator $\Delta : C^k(\mathrm{R}^n) \to C^{k-2}(\mathrm{R}^n)$, or more generally an operator $\Delta : C^k(\Omega) \to C^{k-2}(\Omega)$ for any open set Ω.

Motivation

Diffusion: In the physical theory of diffusion, the Laplace operator (via Laplace's equation) arises naturally in the mathematical description of equilibrium. Specifically, if u is the density at equilibrium of some quantity such as a chemical concentration, then the net flux of u through the boundary of any smooth region V is zero, provided there is no source or sink within V:

$$\int_{\partial V} \nabla u \cdot \mathrm{n} \, dS = 0,$$

where n is the outward unit normal to the boundary of V. By the divergence theorem,

$$\int_V \operatorname{div} \nabla u \, dV = \int_{\partial V} \nabla u \cdot \mathrm{n} \, dS = 0.$$

Since this holds for all smooth regions V, it can be shown that this implies

$$\operatorname{div} \nabla u = \Delta u = 0.$$

The left-hand side of this equation is the Laplace operator. The Laplace operator itself has a physical interpretation for non-equilibrium diffusion as the extent to which a point represents a source or sink of chemical concentration, in a sense made precise by the diffusion equation.

Density Associated to a Potential

If φ denotes the electrostatic potential associated to a charge distribution q, then the charge distribution itself is given by the Laplacian of φ:

$$q = \Delta \varphi.$$

This is a consequence of Gauss's law. Indeed, if V is any smooth region, then by Gauss's law the flux of the electrostatic field E is equal to the charge enclosed (in appropriate units):

$$\int_{\partial V} \mathrm{E}\cdot\mathrm{n}\,dS = \int_{\partial V} \nabla\varphi\cdot\mathrm{n}\,dS = \int_V q\,dV,$$

where the first equality uses the fact that the electrostatic field is the gradient of the electrostatic potential. The divergence theorem now gives

$$\int_V \Delta\varphi\,dV = \int_V q\,dV,$$

and since this holds for all regions V, (1) follows.

The same approach implies that the Laplacian of the gravitational potential is the mass distribution. Often the charge (or mass) distribution are given, and the associated potential is unknown. Finding the potential function subject to suitable boundary conditions is equivalent to solving Poisson's equation.

Energy minimization

Another motivation for the Laplacian appearing in physics is that solutions to $\Delta f = 0$ in a region U are functions that make the Dirichlet energy functional stationary:

$$E(f) = \frac{1}{2}\int_U \|\nabla f\|^2\,dx.$$

To see this, suppose $f : U \to \mathbb{R}$ is a function, and $u : U \to \mathbb{R}$ is a function that vanishes on the boundary of U. Then

$$\frac{d}{d\varepsilon}\Big|_{\varepsilon=0} E(f+\varepsilon u) = \int_U \nabla f\cdot\nabla u\,dx = -\int_U u\Delta f\,dx$$

where the last equality follows using Green's first identity. This calculation shows that if $\Delta f = 0$, then E is stationary around f. Conversely, if E is stationary around f, then $\Delta f = 0$ by the fundamental lemma of calculus of variations.

Coordinate Expressions

Two Dimensions: The Laplace operator in two dimensions is given by

$$\Delta f = \frac{\partial^2 f}{\partial x^2} + \frac{\partial^2 f}{\partial y^2}$$

where x and y are the standard Cartesian coordinates of the xy-plane.

In polar coordinates,

$$\begin{aligned}\Delta f &= \frac{1}{r}\frac{\partial}{\partial r}\left(r\frac{\partial f}{\partial r}\right)+\frac{1}{r^2}\frac{\partial^2 f}{\partial \theta^2}\\ &= \frac{1}{r}\frac{\partial f}{\partial r}+\frac{\partial^2 f}{\partial r^2}+\frac{1}{r^2}\frac{\partial^2 f}{\partial \theta^2}.\end{aligned}$$

Three Dimensions

In three dimensions, it is common to work with the Laplacian in a variety of different coordinate systems.

In Cartesian coordinates,

$$\Delta f = \frac{\partial^2 f}{\partial x^2}+\frac{\partial^2 f}{\partial y^2}+\frac{\partial^2 f}{\partial z^2}.$$

In cylindrical coordinates,

$$\Delta f = \frac{1}{\rho}\frac{\partial}{\partial \rho}\left(\rho\frac{\partial f}{\partial \rho}\right)+\frac{1}{\rho^2}\frac{\partial^2 f}{\partial \phi^2}+\frac{\partial^2 f}{\partial z^2}.$$

In spherical coordinates:

$$\Delta f = \frac{1}{r^2}\frac{\partial}{\partial r}\left(r^2\frac{\partial f}{\partial r}\right)+\frac{1}{r^2\sin\theta}\frac{\partial}{\partial \theta}\left(\sin\theta\frac{\partial f}{\partial \theta}\right)+\frac{1}{r^2\sin^2\theta}\frac{\partial^2 f}{\partial \phi^2}.$$

or:

$$\Delta f = \frac{1}{r}\frac{\partial^2}{\partial r^2}(rf)+\frac{1}{r^2\sin\theta}\frac{\partial}{\partial \theta}\left(\sin\theta\frac{\partial f}{\partial \theta}\right)+\frac{1}{r^2\sin^2\theta}\frac{\partial^2 f}{\partial \varphi^2}.$$

(here φ represents the azimuthal angle and θ the zenith angle or co-latitude).

In general curvilinear coordinates (ξ^1, ξ^2, ξ^3):

$$\nabla^2 = \nabla\xi^m\cdot\nabla\xi^n\frac{\partial^2}{\partial\xi^m\partial\xi^n}+\nabla^2\xi^m\frac{\partial}{\partial\xi^m},$$

where summation over the repeated indices is implied.

N *Dimensions*

In spherical coordinates in N dimensions, with the parametrization $x = r\theta \in \mathrm{R}^N$ with r representing a positive real radius and θ an element of the unit sphere S^{N-1},

$$\Delta f = \frac{\partial^2 f}{\partial r^2} + \frac{N-1}{r}\frac{\partial f}{\partial r} + \frac{1}{r^2}\Delta_{S^{N-1}} f$$

where $\Delta_{S^{N-1}}$ is the Laplace–Beltrami operator on the (N–1)-sphere, known as the spherical Laplacian. The two radial terms can be equivalently rewritten as

$$\frac{1}{r^{N-1}}\frac{\partial}{\partial r}\left(r^{N-1}\frac{\partial f}{\partial r}\right).$$

As a consequence, the spherical Laplacian of a function defined on $S^{N-1} \subset \mathrm{R}^N$ can be computed as the ordinary Laplacian of the function extended to $\mathrm{R}^N \setminus \{0\}$ so that it is constant along rays, i.e., homogeneous of degree zero.

Spectral Theory

The spectrum of the Laplace operator consists of all eigenvalues λ for which there is a corresponding eigenfunction f with

$$-\Delta f = \lambda f.$$

This is known as the Helmholtz equation. If Ω is a bounded domain in R^n then the eigenfunctions of the Laplacian are an orthonormal basis for the Hilbert space $L^2(\Omega)$. This result essentially follows from the spectral theorem on compact self-adjoint operators, applied to the inverse of the Laplacian (which is compact, by the Poincaré inequality and Kondrakov embedding theorem). It can also be shown that the eigenfunctions are infinitely differentiable functions. More generally, these results hold for the Laplace–Beltrami operator on any compact Riemannian manifold with boundary, or indeed for the Dirichlet eigenvalue problem of any elliptic operator with smooth coefficients on a bounded domain. When Ω is the n-sphere, the eigenfunctions of the Laplacian are the well-known spherical harmonics.

Laplace–Beltrami Operator

In differential geometry, the Laplace operator, named after Pierre-Simon Laplace, can be generalized to operate on functions defined on surfaces in Euclidean space and, more generally, on Riemannian and pseudo-Riemannian manifolds. This more general operator goes by the name Laplace–Beltrami operator, after Laplace and Eugenio Beltrami. Like the Laplacian, the Laplace–Beltrami operator is defined as the divergence of the gradient, and is a linear operator taking functions into functions. The operator can be extended to operate on tensors as the divergence of the covariant derivative. Alternatively, the operator can be generalized to operate on differential forms using the divergence and exterior derivative. The resulting operator is called the Laplace–de Rham operator (named after Georges de Rham).

The Laplace–Beltrami operator, like the Laplacian, is the divergence of the gradient:

$$\Delta f = \operatorname{div} \operatorname{grad} f.$$

An explicit formula in local coordinates is possible.

Suppose first that M is an oriented Riemannian manifold. The orientation allows one to specify a definite volume form on M, given in an oriented coordinate system x^i by

$$\operatorname{vol}_n := \sqrt{|g|}\, dx^1 \wedge \ldots \wedge dx^n$$

where the dx^i are the 1-forms forming the dual basis to the basis vectors

$$\partial_i := \frac{\partial}{\partial x^i}$$

and $\wedge$ is the wedge product. Here $|g| := |\det(g_{ij})|$ is the absolute value of the determinant of the metric tensor g_{ij}. The divergence div X of a vector field X on the manifold is then defined as the scalar function with the property

$$(\operatorname{div} X) \operatorname{vol}_n := L_X \operatorname{vol}_n$$

where L_X is the Lie derivative along the vector field X. In local coordinates, one obtains

$$\operatorname{div} X = \frac{1}{\sqrt{|g|}} \partial_i \left(\sqrt{|g|} X^i \right)$$

where the Einstein notation is implied, so that the repeated index i is summed over. The gradient of a scalar function f is the vector field grad f that may be defined through the inner product $\langle \cdot, \cdot \rangle$ on the manifold, as

$$\langle \operatorname{grad} f(x), v_x \rangle = df(x)(v_x)$$

for all vectors v_x anchored at point x in the tangent space T_xM of the manifold at point x. Here, df is the exterior derivative of the function f; it is a 1-form taking argument v_x. In local coordinates, one has

$$(\text{grad}f)^i = \partial^i f = g^{ij}\partial_j f$$

where g^{ij} are the components of the inverse of the metric tensor, so that $g^{ij}g_{jk} = \delta^i_k$ with δ^i_k the Kronecker delta.

Combining the definitions of the gradient and divergence, the formula for the Laplace–Beltrami operator Δ applied to a scalar function f is, in local coordinates

$$\Delta f = \text{div grad} f = \frac{1}{\sqrt{|g|}}\partial_i\left(\sqrt{|g|}g^{ij}\partial_j f\right).$$

If M is not oriented, then the above calculation carries through exactly as presented, except that the volume form must instead be replaced by a volume element (a density rather than a form). Neither the gradient nor the divergence actually depends on the choice of orientation, and so the Laplace–Beltrami operator itself does not depend on this additional structure.

Formal Self-adjointness

The exterior derivative d and "div are formal adjoints, in the sense that for f a compactly supported function

$$\int_M df(X)\,\text{vol}_n = -\int_M f\text{div}X\,\text{vol}_n \quad \text{(proof)}$$

where the last equality is an application of Stokes' theorem. Dualizing gives

$$\int_M f\Delta h\text{vol}_n = -\int_M \langle df, dh\rangle\text{vol}_n$$

for all compactly supported functions f and h. Conversely, (*2*) characterizes Δ completely, in the sense that it is the only operator with this property.

As a consequence, the Laplace–Beltrami operator is negative and formally self-adjoint, meaning that for compactly supported functions f and h,

$$\int_M f\Delta h\,\text{vol}_n = -\int_M \langle df, dh\rangle\,\text{vol}_n = \int_M h\Delta f\,\text{vol}_n.$$

Because the Laplace–Beltrami operator, as defined in this manner, is negative rather than positive, often it is defined with the opposite sign.

Tensor Laplacian

The Laplace–Beltrami operator can be written using the trace of the iterated covariant derivative associated with the Levi-Civita connection. From this perspective, let X_i be a basis of tangent vector fields (not necessarily induced by a coordinate system). Then the Hessian of a function f is the symmetric 2-tensor whose components are given by

$$H(f)_{ij} = H_f(X_i, X_j) = \nabla_{X_i}\nabla_{X_j} f - \nabla_{\nabla_{X_i} X_j} f$$

This is easily seen to transform tensorially, since it is linear in each of the arguments X_i, X_j. The Laplace–Beltrami operator is then the trace of the Hessian with respect to the metric:

$$\Delta f = \sum_{ij} g^{ij} H(f)_{ij}.$$

In abstract indices, the operator is often written

$$\Delta f = \nabla^a \nabla_a f$$

provided it is understood implicitly that this trace is in fact the trace of the Hessian *tensor*.

Because the covariant derivative extends canonically to arbitrary tensors, the Laplace–Beltrami operator defined on a tensor T by

$$\Delta T = g^{ij}\left(\nabla_{X_i}\nabla_{X_j} T - \nabla_{\nabla_{X_i} X_j} T\right)$$

is well-defined.

Laplace–de Rham Operator

More generally, one can define a Laplacian differential operator on sections of the bundle of differential forms on a pseudo-Riemannian manifold. On a Riemannian manifold it is an elliptic operator, while on a Lorentzian manifold it is hyperbolic. The Laplace–de Rham operator is defined by

$$\Delta = \mathrm{d}\delta + \delta\mathrm{d} = (\mathrm{d} + \delta)^2,$$

where d is the exterior derivative or differential and δ is the codifferential, acting as $(-1)^{kn+n+1} * d *$ on k-forms where " is the Hodge star.

When computing Δf for a scalar function f, we have $\delta f = 0$, so that

$$\Delta f = \delta df.$$

Up to an overall sign, The Laplace–de Rham operator is equivalent to the previous definition of the Laplace–Beltrami operator when acting on a scalar function; see the proof for details. On functions, the Laplace–de Rham operator is actually the negative of the Laplace–Beltrami operator, as the conventional normalization of the codifferential assures that the Laplace–de Rham operator is (formally) positive definite, whereas the Laplace–Beltrami operator is typically negative. The sign is a pure convention, however, and both are common in the literature. The Laplace–de Rham operator differs more significantly from the tensor Laplacian restricted to act on skew-symmetric tensors. Apart from the incidental sign, the two operators differ by a Weitzenböck identity that explicitly involves the Ricci curvature tensor.

Examples

Many examples of the Laplace–Beltrami operator can be worked out explicitly.

Euclidean Space: In the usual (orthonormal) Cartesian coordinates x^i on Euclidean space, the metric is reduced to the Kronecker delta, and one therefore has $|g|=1$. Consequently, in this case

$$\Delta f = \frac{1}{\sqrt{|g|}} \partial_i \sqrt{|g|} \partial^i f = \partial_i \partial^i f$$

which is the ordinary Laplacian. In curvilinear coordinates, such as spherical or cylindrical coordinates, one obtains alternative expressions.

Similarly, the Laplace–Beltrami operator corresponding to the Minkowski metric with signature (–+++) is the D'Alembertian.

Spherical Laplacian: The spherical Laplacian is the Laplace–Beltrami operator on the $(n-1)$-sphere with its canonical metric of constant sectional curvature 1. It is convenient to regard the sphere as isometrically embedded into R^n as the unit sphere centred at the origin. Then for a function f on S^{n-1}, the spherical Laplacian is defined by

$$\Delta_{S^{n-1}} f(x) = \Delta f(x / |x|)$$

where $f(x/|x|)$ is the degree zero homogeneous extension of the function f to $\mathrm{R}^n - \{0\}$, and Δ is the Laplacian of the ambient Euclidean space. Concretely, this is implied by the well-known formula for the Euclidean Laplacian in spherical polar coordinates:

$$\Delta f = r^{1-n} \frac{\partial}{\partial r}\left(r^{n-1} \frac{\partial f}{\partial r} \right) + r^{-2} \Delta_{S^{n-1}} f.$$

More generally, one can formulate a similar trick using the normal bundle to define the Laplace–Beltrami operator of any Riemannian manifold isometrically embedded as a hypersurface of Euclidean space.

One can also give an intrinsic description of the Laplace–Beltrami operator on the sphere in a normal coordinate system. Let (t, ξ) be spherical coordinates on the sphere with respect to a particular point p of the sphere (the "north pole"), that is geodesic polar coordinates with respect to p. Here t represents the latitude measurement along a unit speed geodesic from p, and ξ a parameter representing the choice of direction of the geodesic in S^{n-1}. Then the spherical Laplacian has the form:

$$\Delta_{S^{n-1}} f(t, \xi) = \sin^{2-n} t \frac{\partial}{\partial t}\left(\sin^{n-2} t \frac{\partial f}{\partial t} \right) + \sin^{-2} t \Delta_{\xi} f$$

where Δ_{ξ} is the Laplace–Beltrami operator on the ordinary unit $(n - 2)$-sphere.

Hyperbolic Space: A similar technique works in hyperbolic space. Here the hyperbolic space H^{n-1} can be embedded into the n dimensional Minkowski space, a real vector space equipped with the quadratic form

$$q(x) = x_1^2 - x_2^2 - \cdots - x_n^2.$$

Then H^n is the subset of the future null cone in Minkowski space given by

$$H^n = \{x \mid q(x) = 1, x_1 > 1\}.$$

Then

$$\Delta_{H^{n-1}} f = \Box f\left(x / q(x)^{1/2} \right)|_{H^{n-1}}$$

Here $f(x/q(x)^{1/2})$ is the degree zero homogeneous extension of f to the interior of the future null cone and ¡% is the wave operator

$$\Box = \frac{\partial^2}{\partial x_1^2} - \cdots - \frac{\partial^2}{\partial x_n^2}.$$

The operator can also be written in polar coordinates. Let (t, ξ) be spherical coordinates on the sphere with respect to a particular point p of H^{n-1} (say, the centre of the Poincaré disc). Here t represents the hyperbolic distance from p and ξ a parameter representing the choice of direction of the geodesic in S^{n-1}. Then the spherical Laplacian has the form:

$$\Delta_{H^{n-1}} f(t,\xi) = \sinh^{2-n} t \frac{\partial}{\partial t}\left(\sinh^{n-2} t \frac{\partial f}{\partial t}\right) + \sinh^{-2} t \Delta_\xi f$$

where Δ_ξ is the Laplace–Beltrami operator on the ordinary unit $(n-2)$-sphere.

Total Derivative

In the mathematical field of differential calculus, the term total derivative has a number of closely related meanings.

- The total derivative (full derivative) of a function f, of several variables, e.g., t, x, y, etc., with respect to one of its input variables, e.g., t, is different from its partial derivative (∂). Calculation of the total derivative of f with respect to t does not assume that the other arguments are constant while t varies; instead, it allows the other arguments to depend on t. The total derivative adds in these *indirect dependencies* to find the overall dependency of f on t. For example, the total derivative of $f(t,x,y)$ with respect to t is

$$\frac{\mathrm{d}f}{\mathrm{d}t} = \frac{\partial f}{\partial t}\frac{\mathrm{d}t}{\mathrm{d}t} + \frac{\partial f}{\partial x}\frac{\mathrm{d}x}{\mathrm{d}t} + \frac{\partial f}{\partial y}\frac{\mathrm{d}y}{\mathrm{d}t}.$$

Which simplifies to

$$\frac{\mathrm{d}f}{\mathrm{d}t} = \frac{\partial f}{\partial t} + \frac{\partial f}{\partial x}\frac{\mathrm{d}x}{\mathrm{d}t} + \frac{\partial f}{\partial y}\frac{\mathrm{d}y}{\mathrm{d}t}.$$

Consider multiplying both sides of the equation by the differential dt:

$$\mathrm{d}f = \frac{\partial f}{\partial t}\mathrm{d}t + \frac{\partial f}{\partial x}\mathrm{d}x + \frac{\partial f}{\partial y}\mathrm{d}y.$$

The result will be the differential change $d\,f$ in the function f. Because f depends on t, some of that change will be due to the partial derivative of f with t respect to . However, some of that change will also be due to the partial derivatives of f with respect to the variables x and y. So, the differential dt is applied to the total derivatives of and to find differentials dx and dy, which can then be used to find the contribution to dt.

- It refers to a differential operator such as

  $$\frac{\mathrm{d}}{\mathrm{d}x} = \frac{\partial}{\partial x} + \sum_{j=1}^{k} \frac{\mathrm{d}y_j}{\mathrm{d}x}\frac{\partial}{\partial y_j},$$

  which computes the total derivative of a function (with respect to x in this case).

- It refers to the (total) differential $\mathrm{d}f$ of a function, either in the traditional language of infinitesimals or the modern language of differential forms.

- A differential of the form

  $$\sum_{j=1}^{k} f_j(x_1,\ldots,x_k)\,\mathrm{d}x_j$$

  is called a total differential or an exact differential if it is the differential of a function. Again this can be interpreted infinitesimally, or by using differential forms and the exterior derivative.

- It is another name for the derivative as a linear map, i.e., if f is a differentiable function from R^n to R^m, then the (total) derivative (or differential) of f at $x \in \mathrm{R}^n$ is the linear map from R^n to R^m whose matrix is the Jacobian matrix of f at x.

- It is a synonym for the gradient, which is essentially the derivative of a function from R^n to R.

It is sometimes used as a synonym for the material derivative, $\frac{Du}{Dt}$, in fluid mechanics.

Differentiation with Indirect Dependencies

Suppose that f is a function of two variables, x and y. Normally these variables are assumed to be independent. However, in some situations they may be dependent on each other. For example y could be a function of x, constraining the domain of f to a curve in R^2. In this case the partial derivative of f with respect to x does not give the true rate of change of f with respect to changing x because changing x necessarily changes y. The total derivative takes such dependencies into account.

For example, suppose:

$$f(x,y) = xy.$$

The rate of change of f with respect to x is usually the partial derivative of f with respect to x; in this case,

$$\frac{\partial f}{\partial x} = y.$$

However, if y depends on x, the partial derivative does not give the true rate of change of f as x changes because it holds y fixed.

Suppose we are constrained to the line

$$y = x;$$

then

$$f(x,y) = f(x,x) = x^2.$$

In that case, the total derivative of f with respect to x is

$$\frac{\mathrm{d}f}{\mathrm{d}x} = 2x.$$

Instead of immediately substituting for y in terms of x, this can be found equivalently using the chain rule:

$$\frac{\mathrm{d}f}{\mathrm{d}x} = \frac{\partial f}{\partial x} + \frac{\partial f}{\partial y}\frac{\mathrm{d}y}{\mathrm{d}x} = y + x \cdot 1 = x + x = 2x.$$

Notice that this is not equal to the partial derivative:

$$\frac{\mathrm{d}f}{\mathrm{d}x} = 2x \neq \frac{\partial f}{\partial x} = y = x.$$

While one can often perform substitutions to eliminate indirect dependencies, the chain rule provides for a more efficient and general technique. Suppose $M(t, p_1, ..., p_n)$ is a function of time t and n variables p_i which themselves depend on time. Then, the total time derivative of M is

$$\frac{\mathrm{d}M}{\mathrm{d}t} = \frac{\mathrm{d}}{\mathrm{d}t} M\big(t, p_1(t), \ldots, p_n(t)\big).$$

The chain rule for differentiating a function of several variables implies that

$$\frac{\mathrm{d}M}{\mathrm{d}t} = \frac{\partial M}{\partial t} + \sum_{i=1}^{n} \frac{\partial M}{\partial p_i} \frac{\mathrm{d}p_i}{\mathrm{d}t} = \Big(\frac{\partial}{\partial t} + \sum_{i=1}^{n} \frac{\mathrm{d}p_i}{\mathrm{d}t} \frac{\partial}{\partial p_i}\Big)(M).$$

This expression is often used in physics for a gauge transformation of the Lagrangian, as two Lagrangians that differ only by the total time derivative of a function of time and the n generalized coordinates lead to the same equations of motion. An interesting example concerns the resolution of causality concerning the Wheeler-Feynman time-symmetric theory. The operator in brackets (in the final expression) is also called the total derivative operator (with respect to t).

For example, the total derivative of $f(x(t), y(t))$ is

$$\frac{\mathrm{d}f}{\mathrm{d}t} = \frac{\partial f}{\partial x} \frac{\mathrm{d}x}{\mathrm{d}t} + \frac{\partial f}{\partial y} \frac{\mathrm{d}y}{\mathrm{d}t}.$$

Here there is no $\partial f / \partial t$ term since f itself does not depend on the independent variable t directly.

The Total Derivative via Differentials

Differentials provide a simple way to understand the total derivative. For instance, suppose $M(t, p_1, \ldots, p_n)$ is a function of time t and n variables p_i as in the previous section. Then, the differential of M is

$$\mathrm{d}M = \frac{\partial M}{\partial t} \mathrm{d}t + \sum_{i=1}^{n} \frac{\partial M}{\partial p_i} \mathrm{d}p_i.$$

This expression is often interpreted *heuristically* as a relation between infinitesimals. However, if the variables t and p_j are interpreted as functions, and $M(t, p_1, \ldots, p_n)$ is interpreted to mean the composite of M with these functions, then the above expression makes perfect sense as an equality of differential 1-forms, and is immediate from the chain rule for the exterior derivative. The advantage of this point of view is that it takes into account arbitrary dependencies between the variables.

For example, if $p_1^2 = p_2 p_3$ then $2p_1 \, \mathrm{d}p_1 = p_3 \, \mathrm{d}p_2 + p_2 \, \mathrm{d}p_3$. In particular, if the variables p_j are all functions of t, as in the previous section, then

$$\mathrm{d}M = \frac{\partial M}{\partial t}\mathrm{d}t + \sum_{i=1}^{n} \frac{\partial M}{\partial p_i}\frac{\partial p_i}{\partial t}\mathrm{d}t.$$

Total Differential Equation

A *total differential equation* is a differential equation expressed in terms of total derivatives. Since the exterior derivative is a natural operator, in a sense that can be given a technical meaning, such equations are intrinsic and *geometric*.

Application of the Total Differential to Error Estimation

In measurement, the total differential is used in estimating the error Δf of a function f based on the errors Δx, Δy, ... of the parameters $x, y, ...$ Assuming that the interval is short enough for the change to be approximately linear:

$\Delta f(x) = f'(x) \times \Delta x$

and that all variables are independent, then for all variables,

$$\Delta f = f_x \Delta x + f_y \Delta y + \cdots$$

This is because the derivative f_x with respect to the particular parameter x gives the sensitivity of the function f to a change in x, in particular the error Δx. As they are assumed to be independent, the analysis describes the worst-case scenario. The absolute values of the component errors are used, because after simple computation, the derivative may have a negative sign. From this principle the error rules of summation, multiplication etc. are derived, e.g.:

Let $f(a, b) = a \times b$;

$\Delta f = f_a \Delta a + f_b \Delta b$; evaluating the derivatives

$\Delta f = b\Delta a + a\Delta b$; dividing by f, which is $a \times b$

$\Delta f/f = \Delta a/a + \Delta b/b$

That is to say, in multiplication, the total relative error is the sum of the relative errors of the parameters.

Chapter 5

Partial Derivative

In mathematics, a partial derivative of a function of several variables is its derivative with respect to one of those variables, with the others held constant (as opposed to the total derivative, in which all variables are allowed to vary). Partial derivatives are used in vector calculus and differential geometry. The partial derivative of a function f with respect to the variable x is variously denoted by

$$f'_x, f_{,x}, \partial_x f, \text{ or } \frac{\partial f}{\partial x}$$

The partial-derivative symbol is ∂. One of the first known uses of the symbol in mathematics is by Marquis de Condorcet from 1770, who used it for partial differences. The modern partial derivative notation is by Adrien-Marie Legendre (1786), though he later abandoned it; Carl Gustav Jacob Jacobi re-introduced the symbol in 1841. Suppose that f is a function of more than one variable. For instance,

$$z = f(x,y) = x^2 + xy + y^2.$$

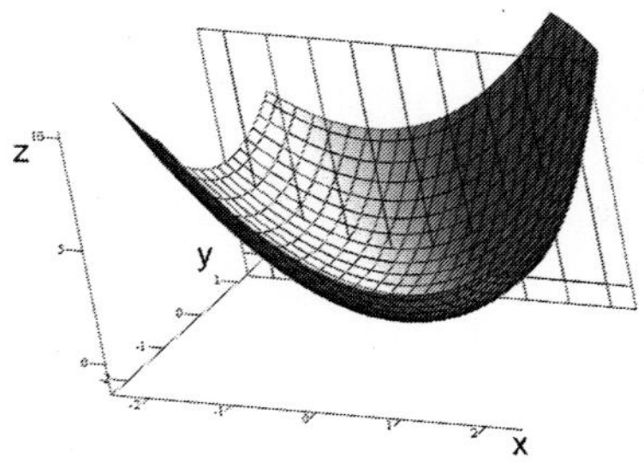

Figure: *A graph of $z = x^2 + xy + y^2$. For the partial derivative at (1, 1, 3) that leaves* y *constant, the corresponding tangent line is parallel to the* xz*-plane.*

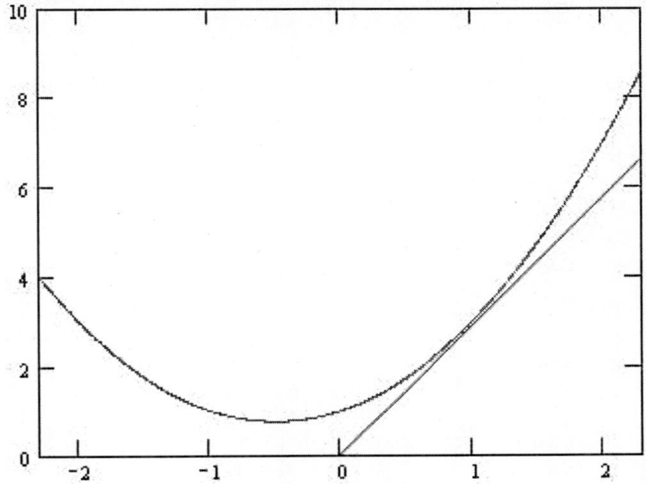

Figure: *A slice of the graph above at y= 1*

The graph of this function defines a surface in Euclidean space. To every point on this surface, there is an infinite number of tangent lines. Partial differentiation is the act of choosing one of these lines and finding its slope.

Usually, the lines of most interest are those that are parallel to the *xz*-plane, and those that are parallel to the *yz*-plane (which result from holding either y or x constant, respectively.)

To find the slope of the line tangent to the function at P(1, 1, 3) that is parallel to the *xz*-plane, the *y* variable is treated as constant. The graph and this plane are shown on the right. On the graph below it, we see the way the function looks on the plane $y = 1$. By finding the derivative of the equation while assuming that y is a constant, the slope of f at the point (x, y, z) is found to be:

$$\frac{\partial z}{\partial x} = 2x + y$$

So at (1, 1, 3), by substitution, the slope is 3. Therefore

$$\frac{\partial z}{\partial x} = 3$$

at the point. (1, 1, 3). That is, the partial derivative of z with respect to x at (1, 1, 3) is 3.

Definition

Basic Definition: The function f can be reinterpreted as a family of functions of one variable indexed by the other variables:

$$f(x,y) = f_x(y) = x^2 + xy + y^2.$$

In other words, every value of x defines a function, denoted f_x, which is a function of one variable. That is,

$$f_x(y) = x^2 + xy + y^2.$$

Once a value of x is chosen, say a, then $f(x,y)$ determines a function f_a which sends y to $a^2 + ay + y^2$:

$$f_a(y) = a^2 + ay + y^2.$$

In this expression, a is a *constant*, not a *variable*, so f_a is a function of only one real variable, that being y. Consequently, the definition of the derivative for a function of one variable applies:

$$f_{a'}(y) = a + 2y.$$

The above procedure can be performed for any choice of a. Assembling the derivatives together into a function gives a function which describes the variation of f in the y direction:

$$\frac{\partial f}{\partial y}(x,y) = x + 2y.$$

This is the partial derivative of f with respect to y. Here ∂ is a rounded d called the partial derivative symbol. To distinguish it from the letter d, ∂ is sometimes pronounced "del" or "partial" instead of "dee".

In general, the partial derivative of a function $f(x_1,...,x_n)$ in the direction x_i at the point $(a_1,...,a_n)$ is defined to be:

$$\frac{\partial f}{\partial x_i}(a_1,\ldots,a_n) = \lim_{h\to 0}\frac{f(a_1,\ldots,a_i+h,\ldots,a_n) - f(a_1,\ldots,a_i,\ldots,a_n)}{h}.$$

In the above difference quotient, all the variables except x_i are held fixed. That choice of fixed values determines a function of one variable

$f_{a_1,\ldots,a_{i-1},a_{i+1},\ldots,a_n}(x_i) = f(a_1,\ldots,a_{i-1},x_i,a_{i+1},\ldots,a_n)$, and by definition,

$$\frac{df_{a_1,\ldots,a_{i-1},a_{i+1},\ldots,a_n}}{dx_i}(x_i) = \frac{\partial f}{\partial x_i}(a_1,\ldots,a_n).$$

In other words, the different choices of a index a family of one-variable functions just as in the example above. This expression also shows that the computation of partial derivatives reduces to the computation of one-variable derivatives.

An important example of a function of several variables is the case of a scalar-valued function $f(x_1,...x_n)$ on a domain in Euclidean space R^n (e.g., on R^2 or R^3). In this case f has a partial derivative $\partial f/\partial x_j$ with respect to each variable x_j. At the point a, these partial derivatives define the vector

$$\nabla f(a) = \left(\frac{\partial f}{\partial x_1}(a),\ldots,\frac{\partial f}{\partial x_n}(a)\right).$$

This vector is called the gradient of f at a. If f is differentiable at every point in some domain, then the gradient is a vector-valued function ∇f which takes the point a to the vector $\nabla f(a)$. Consequently, the gradient produces a vector field.

A common abuse of notation is to define the del operator (∇) as follows in three-dimensional Euclidean space R^3 with unit vectors $\hat{\mathrm{i}},\hat{\mathrm{j}},\hat{\mathrm{k}}$:

$$\nabla = [\frac{\partial}{\partial x}]\hat{\mathrm{i}} + [\frac{\partial}{\partial y}]\hat{\mathrm{j}} + [\frac{\partial}{\partial z}]\hat{\mathrm{k}}$$

Or, more generally, for n-dimensional Euclidean space R^n with coordinates $(x_1, x_2, x_3,\ldots,x_n)$ and unit vectors $(\hat{e}_1,\hat{e}_2,\hat{e}_3,\ldots,\hat{e}_n)$:

$$\nabla = \sum_{j=1}^{n}[\frac{\partial}{\partial x_j}]\hat{e}_j = [\frac{\partial}{\partial x_1}]\hat{e}_1 + [\frac{\partial}{\partial x_2}]\hat{e}_2 + [\frac{\partial}{\partial x_3}]\hat{e}_3 + \ldots + [\frac{\partial}{\partial x_n}]\hat{e}_n$$

Formal Definition

Like ordinary derivatives, the partial derivative is defined as a limit. Let U be an open subset of R^n and $f: U \rightarrow R$ a function. The partial derivative of f at the point $a = (a_1, ..., a_n)$ " U with respect to the i-th variable a_i is defined as

$$\frac{\partial}{\partial a_i}f(\mathrm{a}) = \lim_{h\to 0}\frac{f(a_1,\ldots,a_{i-1},a_i+h,a_{i+1},\ldots,a_n) - f(a_1,\ldots,a_i,\ldots,a_n)}{h}$$

Even if all partial derivatives $\partial f/\partial a_i(a)$ exist at a given point a, the function need not be continuous there. However, if all partial derivatives exist in a neighbourhood of a and are continuous there, then f is totally differentiable in that neighbourhood and the total derivative is continuous. In this case, it is said that f is a C^1 function. This can be used to generalize for vector valued functions ($f : U \to R^m$) by carefully using a componentwise argument.

The partial derivative $\frac{\partial f}{\partial x}$ can be seen as another function defined on U and can again be partially differentiated. If all mixed second order partial derivatives are continuous at a point (or on a set), f is termed a C^2 function at that point (or on that set); in this case, the partial derivatives can be exchanged by Clairaut's theorem:

$$\frac{\partial^2 f}{\partial x_i \partial x_j} = \frac{\partial^2 f}{\partial x_j \partial x_i}.$$

Examples

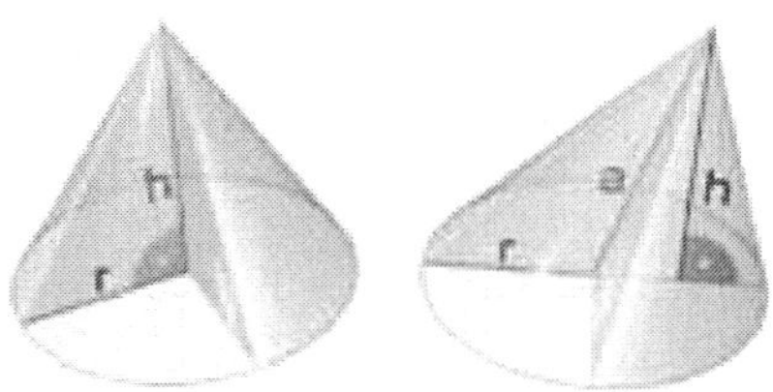

Figure: *The volume of a cone depends on height and radius*

The volume V of a cone depends on the cone's height h and its radius r according to the formula

$$V(r,h) = \frac{\pi r^2 h}{3}.$$

The partial derivative of V with respect to r is

$$\frac{\partial V}{\partial r} = \frac{2\pi r h}{3},$$

which represents the rate with which a cone's volume changes if its radius is varied and its height is kept constant. The partial derivative with respect to h is

$$\frac{\partial V}{\partial h} = \frac{\pi r^2}{3},$$

which represents the rate with which the volume changes if its height is varied and its radius is kept constant.

By contrast, the *total* derivative of V with respect to r and h are respectively

$$\frac{dV}{dr} = \overbrace{\frac{2\pi rh}{3}}^{\frac{\partial V}{\partial r}} + \overbrace{\frac{\pi r^2}{3}}^{\frac{\partial V}{\partial h}} \frac{dh}{dr}$$

and

$$\frac{dV}{dh} = \overbrace{\frac{\pi r^2}{3}}^{\frac{\partial V}{\partial h}} + \overbrace{\frac{2\pi rh}{3}}^{\frac{\partial V}{\partial r}} \frac{dr}{dh}$$

The difference between the total and partial derivative is the elimination of indirect dependencies between variables in partial derivatives.

If (for some arbitrary reason) the cone's proportions have to stay the same, and the height and radius are in a fixed ratio k,

$$k = \frac{h}{r} = \frac{dh}{dr}.$$

This gives the total derivative with respect to r:

$$\frac{dV}{dr} = \frac{2\pi rh}{3} + \frac{\pi r^2}{3}k$$

Which simplifies to:

$$\frac{dV}{dr} = k\pi r^2$$

Similarly, the total derivative with respect to h is:

$$\frac{dV}{dh} = \pi r^2$$

Equations involving an unknown function's partial derivatives are called partial differential equations and are common in physics, engineering, and other sciences and applied disciplines.

Notation

For the following examples, let f be a function in x, y and z.

First-order partial derivatives:

$$\frac{\partial f}{\partial x} = f_x = \partial_x f.$$

Second-order partial derivatives:

$$\frac{\partial^2 f}{\partial x^2} = f_{xx} = \partial_{xx} f.$$

Second-order mixed derivatives:

$$\frac{\partial^2 f}{\partial y \partial x} = \frac{\partial}{\partial y}\left(\frac{\partial f}{\partial x}\right) = f_{xy} = \partial_{yx} f.$$

Higher-order partial and mixed derivatives:

$$\frac{\partial^{i+j+k} f}{\partial x^i \, \partial y^j \, \partial z^k} = f^{(i,j,k)}.$$

When dealing with functions of multiple variables, some of these variables may be related to each other, and it may be necessary to specify explicitly which variables are being held constant. In fields such as statistical mechanics, the partial derivative of f with respect to x, holding y and z constant, is often expressed as

$$\left(\frac{\partial f}{\partial x}\right)_{y,z}.$$

Antiderivative Analogue

There is a concept for partial derivatives that is analogous to antiderivatives for regular derivatives. Given a partial derivative, it allows for the partial recovery of the original function.

Consider the example of $\frac{\partial z}{\partial x} = 2x + y$. The "partial" integral can be taken with respect to x (treating y as constant, in a similar manner to partial derivation):

$$z = \int \frac{\partial z}{\partial x} dx = x^2 + xy + g(y)$$

Here, the "constant" of integration is no longer a constant, but instead a function of all the variables of the original function except x.

The reason for this is that all the other variables are treated as constant when taking the partial derivative, so any function which does not involve x will disappear when taking the partial derivative, and we have to account for this when we take the antiderivative.

The most general way to represent this is to have the "constant" represent an unknown function of all the other variables.

Thus the set of functions $x^2 + xy + g(y)$, where g is any one-argument function, represents the entire set of functions in variables x,y that could have produced the x-partial derivative $2x+y$.

If all the partial derivatives of a function are known (for example, with the gradient), then the antiderivatives can be matched via the above process to reconstruct the original function up to a constant.

Differentiation Rules

This is a summary of differentiation rules, that is, rules for computing the derivative of a function in calculus.

Elementary Rules of Differentiation

Unless otherwise stated, all functions are functions of real numbers (R) that return real values; although more generally, the formulae below apply wherever they are well defined—including complex numbers (C).

Differentiation is Linear

For any functions f and g and any real numbers a and b the derivative of the function $h(x) = af(x) + bg(x)$ with respect to x is

$$h'(x) = af'(x) + bg'(x).$$

In Leibniz's notation this is written as:

$$\frac{d(af + bg)}{dx} = a\frac{df}{dx} + b\frac{dg}{dx}.$$

Special cases include:

- *The constant division rule*

$$(af)' = af'$$

- *The sum rule*

$$(f+g)' = f' + g'$$

- *The subtraction rule*

$$(f-g)' = f' - g'.$$

The product rule (Leibniz rule)

For the functions f and g, the derivative of the function $h(x) = f(x)\, g(x)$ with respect to x is

$$h'(x) = f'(x)g(x) + f(x)g'(x).$$

In Leibniz's notation this is written

$$\frac{d(fg)}{dx} = \frac{df}{dx}g + f\frac{dg}{dx}.$$

The Chain Rule

The derivative of the function of a function $h(x) = f(g(x))$ with respect to x is

$$h'(x) = f'(g(x))g'(x).$$

In Leibniz's notation this is written as:

$$\frac{dh}{dx} = \frac{df(g(x))}{dg(x)}\frac{dg(x)}{dx}.$$

However, by relaxing the interpretation of h as a function, this is often simply written

$$\frac{dh}{dx} = \frac{dh}{dg}\frac{dg}{dx}.$$

The Inverse Function Rule

If the function f has an inverse function g, meaning that $g(f(x)) = x$ and $f(g(y)) = y$, then

$$g' = \frac{1}{f' \circ g}.$$

In Leibniz notation, this is written as

$$\frac{dx}{dy} = \frac{1}{dy/dx}.$$

Power Rule

In calculus, the power rule is one of the most important differentiation rules. Since differentiation is linear, polynomials can be differentiated using this rule.

$$\frac{d}{dx}x^n = nx^{n-1}, \qquad n \neq 0.$$

The power rule holds for all powers except for the constant value x^0 which is covered by the constant rule. The derivative is just 0rather than $0 \cdot x^{-1}$ which is undefined when $x = 0$.

The inverse of the power rule enables all powers of a variable x except x^{-1} to be integrated. This integral is called Cavalieri's quadrature formula and was first found in a geometric form by Bonaventura Cavalieri for $n \geq 0$. It is considered the first general theorem of calculus to be discovered.

$$\int x^n \, dx = \frac{x^{n+1}}{n+1} + C, \qquad n \neq -1.$$

This is an indefinite integral where C is the arbitrary constant of integration.

The integration of x^{-1} requires a separate rule.

$$\int x^{-1} \, dx = \ln|x| + C,$$

Hence, the derivative of x^{100} is $100x^{99}$ and the integral of x^{100} is $\frac{1}{101}x^{101} + C$.

Historically the power rule was derived as the inverse of Cavalieri's quadrature formula which gave the area under x^n for any integer $n \geq 0$. Nowadays the power rule is derived first and integration considered as its inverse.

For integers $n \geq 1$, the derivative of $f(x) = x^n$ is $f'(x) = nx^{n-1}$, that is,

$$\left(x^n\right)' = nx^{n-1}.$$

The power rule *for integration*

$$\int x^n \, dx = \frac{x^{n+1}}{n+1} + C$$

for $n \geq 0$ is then an easy consequence. One just needs to take the derivative of this equality and use the power rule and linearity of differentiation on the right-hand side.

Proof

To prove the power rule for differentiation, we use the definition of the derivative as a limit. But first, note the factorization for $n \geq 1$:

$$f(x) - f(a) = x^n - a^n = (x-a)(x^{n-1} + ax^{n-2} + \cdots + a^{n-2}x + a^{n-1})$$

Using this, we can see that

$$f'(a) = \lim_{x \to a} \frac{x^n - a^n}{x - a} = \lim_{x \to a} x^{n-1} + ax^{n-2} + \cdots + a^{n-2}x + a^{n-1}$$

Since the division has been eliminated and we have a continuous function, we can freely substitute to find the limit:

$$f'(a) = \lim_{x \to a} x^{n-1} + ax^{n-2} + \cdots + a^{n-2}x + a^{n-1} = a^{n-1} + a^{n-1} + \cdots + a^{n-1} + a^{n-1} = n \cdot a^{n-1}$$

The use of the quotient rule allows the extension of this rule for n as a negative integer, and the use of the laws of exponents and the chain rule allows this rule to be extended to all rational values of n. For an irrational n, a rational approximation is appropriate.

Differentiation of Arbitrary Polynomials

To differentiate arbitrary polynomials, one can use the linearity property of the differential operator to obtain:

$$\left(\sum_{r=0}^{n} a_r x^r \right)' = \sum_{r=0}^{n} \left(a_r x^r \right)' = \sum_{r=0}^{n} a_r \left(x^r \right)' = \sum_{r=0}^{n} r a_r x^{r-1}.$$

Using the linearity of integration and the power rule for integration, one shows in the same way that

$$\int \left(\sum_{k=0}^{n} a_k x^k \right) dx = \sum_{k=0}^{n} \frac{a_k x^{k+1}}{k+1} + C.$$

Generalizations

One can prove that the power rule is valid for any exponent r, that is

$$\left(x^r \right)' = r x^{r-1},$$

as long as x is in the domain of the functions on the left and right hand sides and r is nonzero. Using this formula, together with

$$\int x^{-1}\,dx = \ln|x| + C,$$

one can differentiate and integrate linear combinations of powers of x which are not necessarily polynomials.

Reciprocal Rule

In calculus, the reciprocal rule is a shorthand method of finding the derivative of a function that is the reciprocal of a differentiable function, without using the quotient rule or chain rule.

The reciprocal rule states that the derivative of $1/g(x)$ is given by

$$\frac{d}{dx}\left(\frac{1}{g(x)}\right) = \frac{-g'(x)}{(g(x))^2}$$

where $g(x) \neq 0$.

From the Chain Rule

It is also possible to derive the reciprocal rule from the chain rule, by a process very much like that of the derivation of the quotient rule. One thinks of $\frac{1}{g(x)}$ as being the function $\frac{1}{x}$ composed with the function $g(x)$. The result then follows by application of the chain rule.

Examples

The derivative of $1/(x^3 + 4x)$ is:

$$\frac{d}{dx}\left(\frac{1}{x^3 + 4x}\right) = \frac{-3x^2 - 4}{(x^3 + 4x)^2}.$$

The derivative of $1/\cos(x)$ (when $\cos x \neq 0$) is:

$$\frac{d}{dx}\left(\frac{1}{\cos(x)}\right) = \frac{\sin(x)}{\cos^2(x)} = \frac{1}{\cos(x)}\frac{\sin(x)}{\cos(x)} = \sec(x)\tan(x).$$

Gamma Function

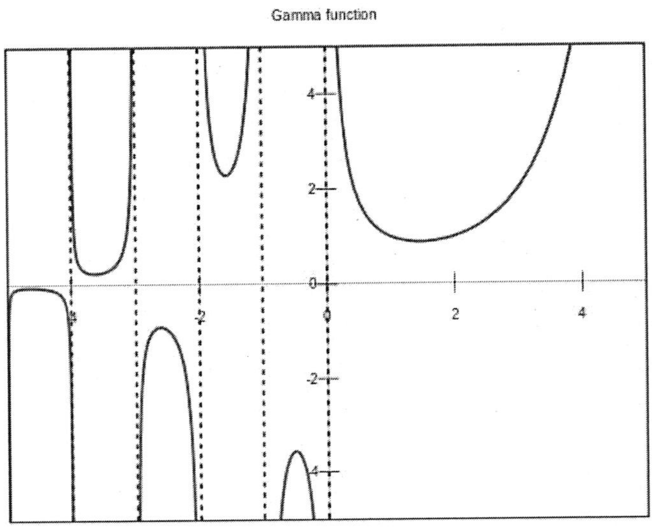

Figure: *The gamma function along part of the real axis*

In mathematics, the gamma function (represented by the capital Greek letter Γ) is an extension of the factorial function, with its argument shifted down by 1, to real and complex numbers. That is, if n is a positive integer:

$$\Gamma(n) = (n-1)!$$

The gamma function is defined for all complex numbers except the negative integers and zero.

For complex numbers with a positive real part, it is defined via an improper integral that converges:

$$\Gamma(z) = \int_0^\infty t^{z-1}e^{-t}\,\mathrm{d}t.$$

This integral function is extended by analytic continuation to all complex numbers except the non-positive integers (where the function has simple poles), yielding the meromorphic function we call the gamma function.

The gamma function is a component in various probability-distribution functions, and as such it is applicable in the fields of probability and statistics, as well as combinatorics.

Motivation

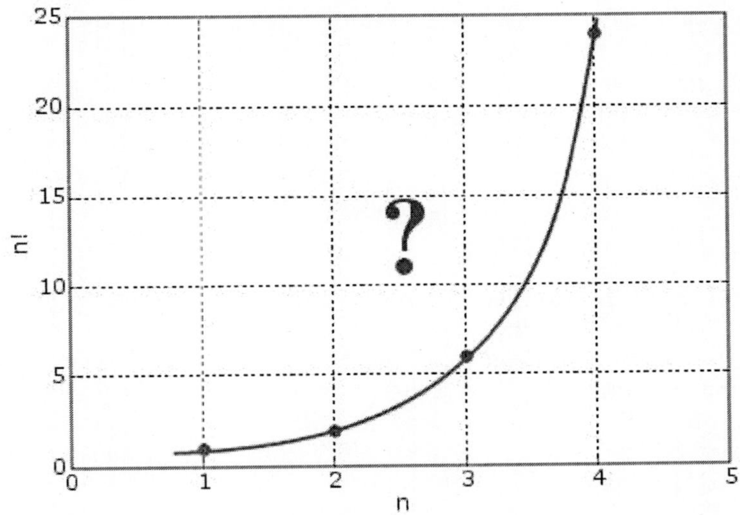

Figure: *It is easy graphically to interpolate the factorial function to non-integer values, but is there a formula that describes the resulting curve?*

The gamma function can be seen as a solution to the following interpolation problem:

"Find a smooth curve that connects the points (x, y) given by $y = (x - 1)!$ at the positive integer values for x."

A plot of the first few factorials makes clear that such a curve can be drawn, but it would be preferable to have a formula that precisely describes the curve, in which the number of operations does not depend on the size of x. The simple formula for the factorial, $n! = 1 \times 2 \times \ldots \times n$, cannot be used directly for fractional values of x since it is only valid when x is a natural number (*i.e.*, a positive integer). There are, relatively speaking, no such simple solutions for factorials; any combination of sums, products, powers, exponential functions, or logarithms with a fixed number of terms will not suffice to express $x!$. Stirling's approximation is asymptotically equal to the factorial function for large values of x. It is possible to find a general formula for factorials using tools such as integrals and limits from calculus. A good solution to this is the gamma function.

There are infinitely many continuous extensions of the factorial to non-integers: infinitely many curves can be drawn through any set of isolated points. The gamma function is the most useful solution in practice, being analytic (except at the non-positive integers), and it can be characterized in several ways. However, it is not the only analytic function which extends the factorial, as adding to it any analytic function which is zero on the positive integers will give another function with that property.

$$\begin{aligned} f(1) &= 1\text{, and} \\ f(x+1) &= xf(x), \end{aligned}$$

for x equal to any positive real number. The Bohr–Mollerup theorem proves that these properties, together with the assumption that f be logarithmically convex (or "superconvex"), uniquely determine f for positive, real inputs. From there, the gamma function can be extended to all real and complex values (except the negative integers and zero) by using the unique analytic continuation of f.

Definition

Main definition:

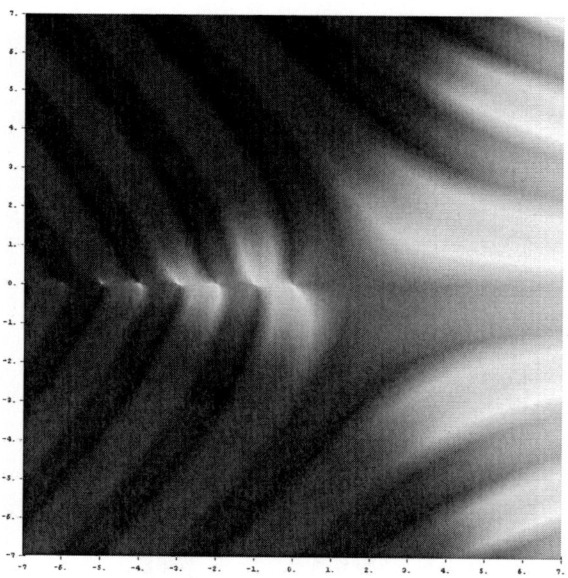

Figure: *The extended version of the gamma function in the complex plane*

The notation $\Gamma(z)$ is due to Legendre. If the real part of the complex number z is positive ($\mathrm{Re}(z) > 0$), then the integral

$$\Gamma(z) = \int_0^\infty t^z e^{-t} \frac{dt}{t}$$

converges absolutely, and is known as the Euler integral of the second kind (the Euler integral of the first kind defines the Beta function). Using integration by parts, we see that the gamma function satisfies the functional equation:

$$\Gamma(z+1) = z\Gamma(z).$$

Combining this with $\Gamma(1) = 1$, we get:

$$\Gamma(n) = 1 \cdot 2 \cdot 3 \cdots (n-1) = (n-1)!$$

for all positive integers n.

The identity $\Gamma(z) = \Gamma(z+1)/z$ can be used (or, yielding the same result, analytic continuation can be used) to extend the integral formulation for $\Gamma(z)$ to a meromorphic function defined for all complex numbers z, except $z = -n$ for integers $n \geq 0$, where the function has simple poles with residue $(-1)^n/n!$.

It is this extended version that is commonly referred to as the gamma function.

Alternative Definitions

The following infinite product definitions for the gamma function, due to Euler and Weierstrass respectively, are valid for all complex numbers z, except the non-positive integers:

$$\Gamma(z) = \lim_{n\to\infty} \frac{n!\, n^z}{z\,(z+1)\cdots(z+n)} = \frac{1}{z}\prod_{n=1}^{\infty} \frac{\left(1+\frac{1}{n}\right)^z}{1+\frac{z}{n}}$$

$$\Gamma(z) = \frac{e^{-\gamma z}}{z}\prod_{n=1}^{\infty}\left(1+\frac{z}{n}\right)^{-1} e^{\frac{z}{n}}$$

where $\gamma \approx 0.577216...$ is the Euler–Mascheroni constant. It is straightforward to show that the Euler definition satisfies the functional equation (1) above.

A somewhat curious parametrization of the gamma function is given in terms of generalized Laguerre polynomials,

$$\Gamma(z) = t^z \sum_{n=0}^{\infty} \frac{L_n^{(z)}(t)}{z+n},$$

which converges for $\mathrm{Re}(z) < 1/2$.

The Gamma Function in the Complex Plane

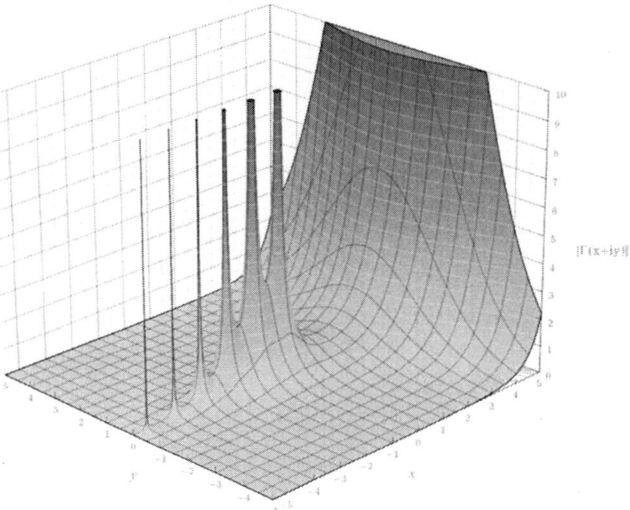

Figure: *The absolute value of the gamma function on the complex plane.*

The behaviour of $\Gamma(z)$ for an increasing positive variable is simple: it grows quickly — faster than an exponential function. Asymptotically as $z \to \infty$, the magnitude of the gamma function is given by Stirling's formula

$$\Gamma(z+1) \sim \sqrt{2\pi z}\left(\frac{z}{e}\right)^z,$$

where the symbol $\sim$ means that the quotient of both sides converges to 1.

The behaviour for nonpositive z is more intricate. Euler's integral does not converge for $z \le 0$, but the function it defines in the positive complex half-plane has a unique analytic continuation to the negative half-plane. One way to find that analytic continuation is to use Euler's integral for positive arguments and extend the domain to negative numbers by repeated application of the recurrence formula,

$$\Gamma(z) = \frac{\Gamma(z+n)}{z(z+1)\cdots(z+n-1)},$$

choosing n such that $z+n$ is positive. The product in the denominator is zero when z equals any of the integers 0, −1, −2,.... Thus, the gamma function must be undefined at those points; it is a meromorphic function with simple poles at the nonpositive integers. The residues of the function at those points are:

$$\operatorname{Res}(\Gamma, -n) = \frac{(-1)^n}{n!}.$$

The gamma function is nonzero everywhere along the real line, although it comes arbitrarily close to zero as $z \to -\infty$. There is in fact no complex number z for which $\Gamma(z) = 0$, and hence the *reciprocal gamma function* $1/\Gamma$ is an entire function, with zeros at $z = 0, -1, -2, \ldots$ The gamma function has a local minimum at $x_{\min} \approx 1.46163$ where it attains the value $\Gamma(x_{\min}) \approx 0.885603$. The gamma function must alternate sign between the poles because the product in the forward recurrence contains an odd number of negative factors if the number of poles between z and $z+n$ is odd, and an even number if the number of poles is even.

Properties

General: Other important functional equations for the gamma function are Euler's reflection formula

$$\Gamma(1-z)\Gamma(z) = \frac{\pi}{\sin(\pi z)},$$

which implies

$$\Gamma(\varepsilon - n) = (-1)^{n-1} \frac{\Gamma(-\varepsilon)\Gamma(1+\varepsilon)}{\Gamma(n+1-\varepsilon)},$$

and the duplication formula

$$\Gamma(z)\Gamma\left(z + \tfrac{1}{2}\right) = 2^{1-2z} \sqrt{\pi}\, \Gamma(2z).$$

The duplication formula is a special case of the multiplication theorem

$$\prod_{k=0}^{m-1} \Gamma\left(z + \frac{k}{m}\right) = (2\pi)^{\frac{m-1}{2}} m^{\frac{1}{2}-mz} \Gamma(mz).$$

A simple but useful property, which can be seen from the limit definition, is:

$$\overline{\Gamma(z)} = \Gamma(\overline{z}).$$

Perhaps the best-known value of the gamma function at a non-integer argument is

$$\Gamma\left(\tfrac{1}{2}\right) = \sqrt{\pi},$$

which can be found by setting $z = 1/2$ in the reflection or duplication formulas, by using the relation to the beta function given below with $x = y = 1/2$, or simply by making the substitution $u = \sqrt{t}$ in the integral definition of the gamma function, resulting in a Gaussian integral. In general, for non-negative integer values of n we have:

$$\Gamma\left(\tfrac{1}{2}+n\right) = \frac{(2n)!}{4^n n!}\sqrt{\pi} = \frac{(2n-1)!!}{2^n}\sqrt{\pi} = \sqrt{\pi}\left[\binom{n-\frac{1}{2}}{n} n!\right]$$

$$\Gamma\left(\tfrac{1}{2}-n\right) = \frac{(-4)^n n!}{(2n)!}\sqrt{\pi} = \frac{(-2)^n}{(2n-1)!!}\sqrt{\pi} = \frac{\sqrt{\pi}}{\binom{-\frac{1}{2}}{n} n!}$$

where $n!!$ denotes the double factorial and, when $n = 0$, $n!! = 1$.

It might be tempting to generalize the result that $\Gamma(1/2) = -\pi$ by looking for a formula for other individual values $\Gamma(r)$ where r is rational. However, these numbers are not known to be expressible by themselves in terms of elementary functions. It has been proved that $\Gamma(n+r)$ is a transcendental number and algebraically independent of π for any integer n and each of the fractions $r = 1/6$, $1/4$, $1/3$, $2/3$, $3/4$, and $5/6$. In general, when computing values of the gamma function, we must settle for numerical approximations.

Another useful limit for asymptotic approximations is:

$$\lim_{n\to\infty} \frac{\Gamma(n+\alpha)}{\Gamma(n)n^{\alpha}} = 1, \qquad \alpha \in \mathrm{R}$$

The derivatives of the gamma function are described in terms of the polygamma function. For example:

$$\Gamma'(z) = \Gamma(z)\psi_0(z).$$

For positive integer m the derivative of gamma function can be calculated as follows (here γ is the Euler–Mascheroni constant):

$$\Gamma'(m+1) = m!\left(-\gamma + \sum_{k=1}^{m} \frac{1}{k}\right).$$

The n-th derivative of the gamma function is:

$$\frac{\mathrm{d}^n}{\mathrm{d}x^n}\Gamma(x) = \int_0^\infty t^{x-1} e^{-t} (\ln t)^n \,\mathrm{d}t.$$

The gamma function has simple poles at $z = -n = 0, -1, -2, -3, \ldots$. The residue there is

$$\mathrm{Res}(\Gamma, -n) = \frac{(-1)^n}{n!}.$$

Moreover, the gamma function has the following Laurent expansion in 1

$$\Gamma(z) = 1 + \sum_{k=1}^{\infty} \frac{\Gamma^{(k)}(1)}{k!}(z-1)^{k-1},$$

valid for $|z| < 1$. In particular

$$\Gamma(z) = \frac{1}{z} - \gamma + \frac{1}{6}\left(3\gamma^2 + \frac{\pi^2}{2}\right)z + O(z^2).$$

The Bohr–Mollerup theorem states that among all functions extending the factorial functions to the positive real numbers, only the gamma function is log-convex, that is, its natural logarithm is convex on the positive real axis.

In a certain sense, the log(Γ)-function is the more natural form therefore. The natural logarithm is not taken because Γ-values are growing so rapidly for increasing real arguments, but because intrinsic attributes become clearer. A striking example is the Taylor series of log(Γ) in 1:

$$\ln\Gamma(z+1) = -\gamma z + \sum_{k=2}^{\infty} \frac{\zeta(k)}{k}(-z)^k \qquad \forall\, |z| < 1$$

with $\zeta(k)$ denoting the Riemann zeta function at k.

Pi Function

An alternative notation which was originally introduced by Gauss and which was sometimes used is the Pi function, which in terms of the gamma function is

$$\Pi(z) = \Gamma(z+1) = z\Gamma(z) = \int_0^\infty e^{-t} t^z \, dt,$$

so that

$$\Pi(n) = n!,$$

for every non-negative integer n.

Using the Pi function the reflection formula takes on the form

$$\Pi(z)\Pi(-z) = \frac{\pi z}{\sin(\pi z)} = \frac{1}{\operatorname{sinc}(z)}$$

where sinc is the normalized sinc function, while the multiplication theorem takes on the form

$$\Pi\left(\frac{z}{m}\right)\Pi\left(\frac{z-1}{m}\right)\cdots\Pi\left(\frac{z-m+1}{m}\right) = (2\pi)^{\frac{m-1}{2}} m^{-z-\frac{1}{2}}\Pi(z).$$

We also sometimes find

$$\pi(z) = \frac{1}{\Pi(z)},$$

which is an entire function, defined for every complex number. That $\pi(z)$ is entire entails it has no poles, so $\Gamma(z)$ has no zeros.

Somewhat interestingly, the Volume of an n-ellipsoid with radii $r_1,\ldots,r_n$ can be expressed as

$$V_n(r_1,\ldots,r_n) = \frac{\pi^{\frac{n}{2}}}{\Pi(\frac{n}{2})}\prod_{k=1}^{n} r_k$$

Relation to Other Functions

- In the first integral above, which defines the gamma function, the limits of integration are fixed. The upper and lower incomplete gamma functions are the functions obtained by allowing the lower or upper (respectively) limit of integration to vary.

- The gamma function is related to the Beta function by the formula

$$\mathrm{B}(x,y)=\frac{\Gamma(x)\,\Gamma(y)}{\Gamma(x+y)}.$$

- The derivative of the logarithm of the gamma function is called the digamma function; higher derivatives are the polygamma functions.
- The analog of the gamma function over a finite field or a finite ring is the Gaussian sums, a type of exponential sum.
- The reciprocal gamma function is an entire function and has been studied as a specific topic.
- The gamma function also shows up in an important relation with the Riemann zeta function, $\zeta(z)$.

$$\pi^{-\frac{z}{2}}\,\Gamma\left(\frac{z}{2}\right)\zeta(z)=\pi^{-\frac{1-z}{2}}\,\Gamma\left(\frac{1-z}{2}\right)\zeta(1-z).$$

And also in the following elegant formula:

$$\zeta(z)\,\Gamma(z)=\int_0^\infty \frac{u^{z-1}}{e^u-1}\,\mathrm{d}u,$$

which is valid only for $\mathrm{Re}(z) > 1$.

The logarithm of the gamma function satisfies the following formula due to Lerch:

$$\log\Gamma(x)=\zeta_{H'}(0,x)-\zeta'(0),$$

where ζ_H is the Hurwitz zeta function, ζ is the Riemann zeta function and the prime (') denotes differentiation in the first variable.

- The gamma function is intimately related to the stretched exponential function. For instance, the moments of that function are

$$\langle\tau^n\rangle \equiv \int_0^\infty \mathrm{d}t\,t^{n-1}\,e^{-\left(\frac{t}{\tau}\right)^\beta}=\frac{\tau^n}{\beta}\Gamma\left(\frac{n}{\beta}\right).$$

Particular Values

Some particular values of the gamma function are:

$$
\begin{aligned}
\Gamma(-\tfrac{3}{2}) &= \tfrac{4}{3}\sqrt{\pi} && \approx 2.363271801207 \\
\Gamma(-\tfrac{1}{2}) &= -2\sqrt{\pi} && \approx -3.544907701811 \\
\Gamma(\tfrac{1}{2}) &= \sqrt{\pi} && \approx 1.772453850905 \\
\Gamma(1) &= 0! && = 1 \\
\Gamma(\tfrac{3}{2}) &= \tfrac{1}{2}\sqrt{\pi} && \approx 0.88622692545 \\
\Gamma(2) &= 1! && = 1 \\
\Gamma(\tfrac{5}{2}) &= \tfrac{3}{4}\sqrt{\pi} && \approx 1.32934038818 \\
\Gamma(3) &= 2! && = 2 \\
\Gamma(\tfrac{7}{2}) &= \tfrac{15}{8}\sqrt{\pi} && \approx 3.32335097045 \\
\Gamma(4) &= 3! && = 6
\end{aligned}
$$

Approximations

Complex values of the gamma function can be computed numerically with arbitrary precision using Stirling's approximation or the Lanczos approximation.

The gamma function can be computed to fixed precision for $\mathrm{Re}(z) \in [1, 2]$ by applying integration by parts to Euler's integral. For any positive number x the gamma function can be written

$$
\begin{aligned}
\Gamma(z) &= \int_0^x e^{-t} t^{z-1} \frac{\mathrm{d}t}{t} + \int_x^\infty e^{-t} t^{z+1} \frac{\mathrm{d}t}{t} \\
&= x^z e^{-x} \sum_{n=0}^{\infty} \frac{x^n}{z(z+1)\cdots(z+n)} + \int_x^\infty e^{-t} t^z \frac{\mathrm{d}t}{t}.
\end{aligned}
$$

When $\mathrm{Re}(z) \in [1, 2]$ and $x \geq 1$, the absolute value of the last integral is smaller than $(x + 1) \geq^x$. By choosing x large enough, this last expression can be made smaller than 2^{-N} for any desired value N. Thus, the gamma function can be evaluated to N bits of precision with the above series.

The only fast algorithm for calculation of the Euler gamma function for any algebraic argument (including rational) was constructed by E.A. Karatsuba.

For arguments that are integer multiples of 1/24 the gamma function can also be evaluated quickly using arithmetic-geometric mean iterations.

Because the Gamma and factorial functions grow so rapidly for moderately large arguments, many computing environments include a function that returns the natural logarithm of the gamma function (often given the name *lngamma* in programming environments or *gammaln* in spreadsheets); this grows much more slowly, and for combinatorial calculations allows adding and subtracting logs instead of multiplying and dividing very large values. The digamma function, which is the derivative of this function, is also commonly seen. In the context of technical and physical applications, e.g. with wave propagation, the functional equation

$$\ln(\Gamma(z)) = \ln(\Gamma(z+1)) - \ln(z)$$

is often used since it allows one to determine function values in one strip of width 1 in z from the neighbouring strip. In particular, starting with a good approximation for a z with large real part one may go step by step down to the desired z. Following an indication of Carl Friedrich Gauss, Rocktaeschel (1922) proposed for $\ln(\Gamma(z))$ an approximation for large $\mathrm{Re}(z)$:

$$\ln(\Gamma(z)) \approx (z-\tfrac{1}{2})\ln(z) - z + \tfrac{1}{2}\ln(2\pi).$$

This can be used to accurately approximate $\ln(\Gamma(z))$ for z with a smaller $\mathrm{Re}(z)$ via (P.E.Böhmer, 1939)

$$\ln(\Gamma(z-m)) = \ln(\Gamma(z)) - \sum_{k=1}^{m} \ln(z-k).$$

A more accurate approximation can be obtained by using more terms from the asymptotic expansions of $\ln(\Gamma(z))$ and $\Gamma(z)$, which are based on Stirling's approximation.

An asymptotic approximation (which has no convergence area for z!) of the gamma function is

$$\Gamma(z) = z^{z-\frac{1}{2}} e^{-z} \sqrt{2\pi}\left(1 + \frac{1}{12z} + \frac{1}{288z^2} - \frac{139}{51840z^3} - \frac{571}{2488320z^4} + O\left(\frac{1}{z^5}\right)\right) \qquad \text{for } |\arg(z)| < \pi$$

and in its "natural" presentation

$$\ln(\Gamma(z)) = z\ln(z) - z - \tfrac{1}{2}\ln\left(\frac{z}{2\pi}\right) + \frac{1}{12z} - \frac{1}{360z^3} + \frac{1}{1260z^5} + O\left(\frac{1}{z^7}\right) \qquad \text{for } |\arg(z)| < \pi$$

where the coefficients of the terms with $k > 1$ of z^{-k+1} in the last expansion are simply

$$\frac{B_k}{k(k-1)}$$

Applications

Opening a random page in an advanced table of formulas, one may be as likely to spot the gamma function as a trigonometric function. One author describes the gamma function as "Arguably, the most common special function, or the least 'special' of them. The other transcendental functions listed below are called 'special' because you could conceivably avoid some of them by staying away from many specialized mathematical topics. On the other hand, the gamma function $y = \Gamma(x)$ is most difficult to avoid."

Integration Problems

The gamma function finds application in such diverse areas as quantum physics, astrophysics and fluid dynamics. The gamma distribution, which is formulated in terms of the gamma function, is used in statistics to model a wide range of processes; for example, the time between occurrences of earthquakes.

The primary reason for the gamma function's usefulness in such contexts is the prevalence of expressions of the type $f(t)e^{-g(t)}$ which describe processes that decay exponentially in time or space. Integrals of such expressions can occasionally be solved in terms of the gamma function when no elementary solution exists. For example, if f is a power function and g is a linear function, a simple change of variables gives the evaluation

$$\int_0^\infty t^b e^{-at}\,dt = \frac{\Gamma(b+1)}{a^{b+1}}.$$

The fact that the integration is performed along the entire positive real line might signify that the gamma function describes the cumulation of a time-dependent process that continues indefinitely, or the value might be the total of a distribution in an infinite space.

It is of course frequently useful to take limits of integration other than 0 and " to describe the cumulation of a finite process, in which case the ordinary gamma function is no longer a solution; the solution is then called an incomplete gamma function. (The ordinary gamma function, obtained by integrating across the entire positive real line, is sometimes called the *complete gamma function* for contrast).

An important category of exponentially decaying functions is that of Gaussian functions

$$ae^{-\frac{(x-b)^2}{c^2}}$$

and integrals thereof, such as the error function. There are many interrelations between these functions and the gamma function; notably, the square root of π we obtained by evaluating Γ(1/2) is the "same" as that found in the normalizing factor of the error function and the normal distribution.

The integrals we have discussed so far involve transcendental functions, but the gamma function also arises from integrals of purely algebraic functions. In particular, the arc lengths of ellipses and of the lemniscate, which are curves defined by algebraic equations, are given by elliptic integrals that in special cases can be evaluated in terms of the gamma function. The gamma function can also be used to calculate "volume" and "area" of n-dimensional hyperspheres.

Another important special case is that of the beta function

$$\mathrm{B}(x,y) = \int_0^1 t^{x-1}(1-t)^{y-1}\,dt = \frac{\Gamma(x)\Gamma(y)}{\Gamma(x+y)}.$$

Calculating Products

The gamma function's ability to and generalize factorial products immediately leads to applications in many areas of mathematics; in combinatorics, and by extension in areas such as probability theory and the calculation of power series. Many expressions involving products of successive integers can be written as some combination of factorials, the most important example perhaps being that of the binomial coefficient

$$\binom{n}{k} = \frac{n!}{k!(n-k)!}.$$

The example of binomial coefficients motivates why the properties of the gamma function when extended to negative numbers are natural. A binomial coefficient gives the number of ways to choose k elements from a set of n elements; if $k > n$, there are of course no ways. If $k > n$, $(n-k)!$ is the factorial of a negative integer and hence infinite if we use the gamma function definition of factorials — dividing by infinity gives the expected value of 0.

We can replace the factorial by a gamma function to extend any such formula to the complex numbers. Generally, this works for any product wherein each factor is a rational function of the index variable, by factoring the rational function into linear expressions. If P and Q are monic polynomials of degree m and n with respective roots $p_1 \ldots p_m$ and $q_1 \ldots q_n$, we have

$$\prod_{i=a}^{b} \frac{P(i)}{Q(i)} = \left(\prod_{j=1}^{m} \frac{\Gamma(b-p_j+1)}{\Gamma(a-p_j)}\right)\left(\prod_{k=1}^{n} \frac{\Gamma(a-q_k)}{\Gamma(b-q_k+1)}\right).$$

If we have a way to calculate the gamma function numerically, it is a breeze to calculate numerical values of such products. The number of gamma functions in the right-hand side depends only on the degree of the polynomials, so it does not matter whether $b-a$ equals 5 or 10^5. Moreover, due to the poles of the gamma function, the equation also holds (in the sense of taking limits) when the left-hand product contain zeros or poles.

By taking limits, certain rational products with infinitely many factors can be evaluated in terms of the gamma function as well. Due to the Weierstrass factorization theorem, analytic functions can be written as infinite products, and these can sometimes be represented as finite products or quotients of the gamma function. We have already seen one striking example: the reflection formula essentially represents the sine function as the product of two gamma functions. Starting from this formula, the exponential function as well as all the trigonometric and hyperbolic functions can be expressed in terms of the gamma function.

More functions yet, including the hypergeometric function and special cases thereof, can be represented by means of complex contour integrals of products and quotients of the gamma function, called Mellin-Barnes integrals.

Analytic Number Theory

An elegant and deep application of the gamma function is in the study of the Riemann zeta function. A fundamental property of the Riemann zeta function is its functional equation:

$$\Gamma\left(\frac{s}{2}\right)\zeta(s)\pi^{-\frac{s}{2}} = \Gamma\left(\frac{1-s}{2}\right)\zeta(1-s)\pi^{-\frac{1-s}{2}}.$$

Among other things, this provides an explicit form for the analytic continuation of the zeta function to a meromorphic function in the complex plane and leads to an immediate proof that the zeta function has infinitely many so-called "trivial" zeros on the real line. Borwein *et al.* call this formula "one of the most beautiful findings in mathematics". Another champion for that title might be

$$\zeta(z)\,\Gamma(z) = \int_0^\infty \frac{t^z}{e^t - 1}\frac{dt}{t}.$$

Both formulas were derived by Bernhard Riemann in his seminal 1859 paper "Über die Anzahl der Primzahlen unter einer gegebenen Grösse" ("On the Number of Prime Numbers less than a Given Quantity"), one of the milestones in the development of analytic number theory — the branch of mathematics that studies prime numbers using the tools of mathematical analysis.

Factorial numbers, considered as discrete objects, are an important concept in classical number theory because they contain many prime factors, but Riemann found a use for their continuous extension that arguably turned out to be even more important.

History

The gamma function has caught the interest of some of the most prominent mathematicians of all time. Its history, notably documented by Philip J. Davis in an article that won him the 1963 Chauvenet Prize, reflects many of the major developments within mathematics since the 18th century. In the words of Davis, "each generation has found something of interest to say about the gamma function. Perhaps the next generation will also."

18th Century: Euler and Stirling

The problem of extending the factorial to non-integer arguments was apparently first considered by Daniel Bernoulli and Christian Goldbach in the 1720s, and was solved at the end of the same decade by Leonhard Euler. Euler gave two different definitions: the first was not his integral but an infinite product,

$$n! = \prod_{k=1}^{\infty} \frac{\left(1+\frac{1}{k}\right)^n}{1+\frac{n}{k}},$$

of which he informed Goldbach in a letter dated October 13, 1729. He wrote to Goldbach again on January 8, 1730, to announce his discovery of the integral representation

$$n! = \int_0^1 (-\ln s)^n \, ds,$$

which is valid for $n > 0$. By the change of variables $t = -\ln s$, this becomes the familiar Euler integral. Euler published his results in the paper "De progressionibus transcendentibus seu quarum termini generales algebraice dari nequeunt" ("On transcendental progressions, that is, those whose general terms cannot be given algebraically"), submitted to the St. Petersburg Academy on November 28, 1729. Euler further discovered some of the gamma function's important functional properties, including the reflection formula.

James Stirling, a contemporary of Euler, also attempted to find a continuous expression for the factorial and came up with what is now known as Stirling's formula. Although Stirling's formula gives a good estimate of $n!$, also for non-integers, it does not provide the exact value. Extensions of his formula that correct the error were given by Stirling himself and by Jacques Philippe Marie Binet.

19th Century: Gauss, Weierstrass and Legendre

Carl Friedrich Gauss rewrote Euler's product as

$$\Gamma(z) = \lim_{m\to\infty} \frac{m^z m!}{z(z+1)(z+2)\cdots(z+m)}$$

and used this formula to discover new properties of the gamma function. Although Euler was a pioneer in the theory of complex variables, he does not appear to have considered the factorial of a complex number, as instead Gauss first did. Gauss also proved the multiplication theorem of the gamma function and investigated the connection between the gamma function and elliptic integrals.

Karl Weierstrass further established the role of the gamma function in complex analysis, starting from yet another product representation,

$$\Gamma(z) = \frac{e^{-\gamma z}}{z} \prod_{k=1}^{\infty} \left(1 + \frac{z}{k}\right)^{-1} e^{\frac{z}{k}},$$

where γ is the Euler–Mascheroni constant. Weierstrass originally wrote his product as one for $1/\Gamma$, in which case it is taken over the function's zeros rather than its poles. Inspired by this result, he proved what is known as the Weierstrass factorization theorem—that any entire function can be written as a product over its zeros in the complex plane; a generalization of the fundamental theorem of algebra.

The name gamma function and the symbol Γ were introduced by Adrien-Marie Legendre around 1811; Legendre also rewrote Euler's integral definition in its modern form. Although the symbol is an upper-case Greek gamma, there is no accepted standard for whether the function name should be written "gamma function" or "Gamma function" (some authors simply write "Γ-function"). The alternative "Pi function" notation $\Pi(z) = z!$ due to Gauss is sometimes encountered in older literature, but Legendre's notation is dominant in modern works.

It is justified to ask why we distinguish between the "ordinary factorial" and the gamma function by using distinct symbols, and particularly why the gamma function should be normalized to $\Gamma(n+1) = n!$ instead of simply using $-\Gamma(n) = n!$". Consider that the notation for exponents, x^n, has been generalized from integers to complex numbers x^z without any change. Legendre's motivation for the normalization does not appear to be known, and has been criticized as cumbersome by some (the 20th-century mathematician Cornelius Lanczos, for example, called it "void of any rationality" and would instead use $z!$). Legendre's normalization does simplify a few formulas, but complicates most others. From a modern point of view, the

Legendre normalization of the Gamma function is the integral of the additive character $\geq^x$ against the multiplicative character x^z with respect to the Haar measure dx/x on the Lie group R^+. Thus this normalization makes it clearer that the Gamma function is a continuous analogue of a Gauss sum.

19th-20th Centuries: Characterizing the Gamma Function

It is somewhat problematic that a large number of definitions have been given for the gamma function. Although they describe the same function, it is not entirely straightforward to prove the equivalence. Stirling never proved that his extended formula corresponds exactly to Euler's gamma function; a proof was first given by Charles Hermite in 1900. Instead of finding a specialized proof for each formula, it would be desirable to have a general method of identifying the gamma function.

One way to prove would be to find a differential equation that characterizes the gamma function. Most special functions in applied mathematics arise as solutions to differential equations, whose solutions are unique. However, the gamma function does not appear to satisfy any simple differential equation. Otto Hölder proved in 1887 that the gamma function at least does not satisfy any *algebraic* differential equation by showing that a solution to such an equation could not satisfy the gamma function's recurrence formula. This result is known as Hölder's theorem.

A definite and generally applicable characterization of the gamma function was not given until 1922. Harald Bohr and Johannes Mollerup then proved what is known as the *Bohr–Mollerup theorem*: that the gamma function is the unique solution to the factorial recurrence relation that is positive and *logarithmically convex* for positive z and whose value at 1 is 1 (a function is logarithmically convex if its logarithm is convex).

The Bohr–Mollerup theorem is useful because it is relatively easy to prove logarithmic convexity for any of the different formulas used to define the gamma function. Taking things further, instead of defining the gamma function by any particular formula, we can choose the conditions of the Bohr–Mollerup theorem as the definition, and then pick any formula we like that satisfies the conditions as a starting point for studying the gamma function. This approach was used by the Bourbaki group.

Reference Tables and Software

Although the gamma function can be calculated virtually as easily as any mathematically simpler function with a modern computer—even with a programmable pocket calculator—this was of course not always the case. Until the mid-20th century, mathematicians relied on hand-made tables; in the case of the gamma function, notably a table computed by Gauss in 1813 and one computed by Legendre in 1825.

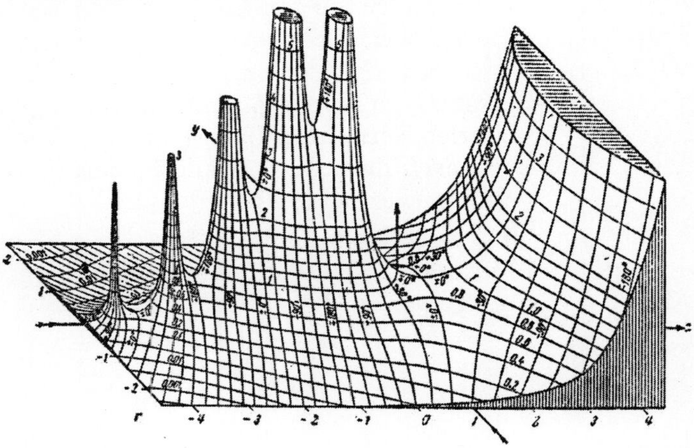

Figure: *A hand-drawn graph of the absolute value of the complex gamma function, from Tables of Higher Functions by Jahnke and Emde.*

Tables of complex values of the gamma function, as well as hand-drawn graphs, were given in *Tables of Higher Functions* by Jahnke and Emde, first published in Germany in 1909. According to Michael Berry, "the publication in J&E of a three-dimensional graph showing the poles of the gamma function in the complex plane acquired an almost iconic status."

There was in fact little practical need for anything but real values of the gamma function until the 1930s, when applications for the complex gamma function were discovered in theoretical physics. As electronic computers became available for the production of tables in the 1950s, several extensive tables for the complex gamma function were published to meet the demand, including a table accurate to 12 decimal places from the U.S. National Bureau of Standards.

Abramowitz and Stegun became the standard reference for this and many other special functions after its publication in 1964.

Double-precision floating-point implementations of the gamma function and its logarithm are now available in most scientific computing software and special functions libraries, for example Matlab, GNU Octave, and the GNU Scientific Library.

The gamma function was also added to the C standard library (math.h).

Arbitrary-precision implementations are available in most computer algebra systems, such as Mathematica and Maple. PARI/GP, MPFR and MPFUN contain free arbitrary-precision implementations.

Riemann Zeta Function

The Riemann zeta function or Euler–Riemann zeta function, $\zeta(s)$, is a function of a complex variable s that analytically continues the sum of the infinite series $\sum_{n=1}^{\infty} \frac{1}{n^s}$, which converges when the real part of s is greater than 1. More general representations of $\zeta(s)$ for all s are given below. The Riemann zeta function plays a pivotal role in analytic number theory and has applications in physics, probability theory, and applied statistics.

This function, as a function of a real argument, was introduced and studied by Leonhard Euler in the first half of the eighteenth century without using complex analysis, which was not available at that time.

Bernhard Riemann in his memoir "On the Number of Primes Less Than a Given Magnitude" published in 1859 extended the Euler definition to a complex variable, proved its meromorphic continuation and functional equation and established a relation between its zeros and the distribution of prime numbers.

The values of the Riemann zeta function at even positive integers were computed by Euler. The first of them, $\zeta(2)$, provides a solution to the Basel problem. In 1979 Apéry proved the irrationality of $\zeta(3)$.

The values at negative integer points, also found by Euler, are rational numbers and play an important role in the theory of modular forms.

Many generalizations of the Riemann zeta function, such as Dirichlet series, Dirichlet L-functions and L-functions, are known.

Definition

The Riemann zeta function $\zeta(s)$ is a function of a complex variable $s = \sigma + it$. (The notation with s, σ, and t is traditionally used in the study of the ζ-function, following Riemann.)

The following infinite series converges for all complex numbers s with real part greater than 1, and defines $\zeta(s)$ in this case:

$$\zeta(s) = \sum_{n=1}^{\infty} n^{-s} = \frac{1}{1^s} + \frac{1}{2^s} + \frac{1}{3^s} + \cdots \qquad \sigma = \Re(s) > 1.$$

The Riemann zeta function is defined as the analytic continuation of the function defined for $\sigma > 1$ by the sum of the preceding series.

Leonhard Euler considered the above series in 1740 for positive integer values of s, and later Chebyshev extended the definition to real $s > 1$.

The above series is a prototypical Dirichlet series that converges absolutely to an analytic function for s such that $\sigma > 1$ and diverges for all other values of s.

Riemann showed that the function defined by the series on the half-plane of convergence can be continued analytically to all complex values $s \neq 1$. For $s = 1$ the series is the harmonic series which diverges to $+\infty$, and

$$\lim_{s \to 1} (s-1)\zeta(s) = 1.$$

Thus the Riemann zeta function is a meromorphic function on the whole complex s-plane, which is holomorphic everywhere except for a simple pole at $s = 1$ with residue 1.

Specific Values

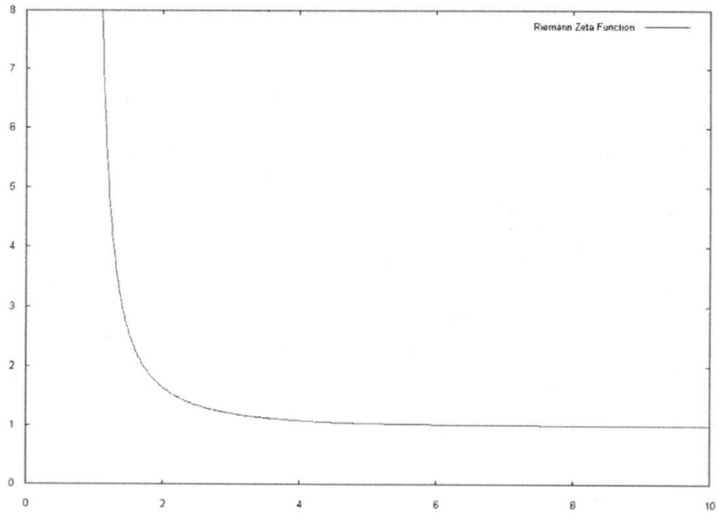

Figure: *Riemann zeta function for real* s > *1*

For any positive even number $2n$,

$$\zeta(2n) = (-1)^{n+1} \frac{B_{2n}(2\pi)^{2n}}{2(2n)!}$$

where B_{2n} is a Bernoulli number; for negative integers, one has

$$\zeta(-n) = -\frac{B_{n+1}}{n+1}$$

for $n \geq 1$, so in particular ζ vanishes at the negative even integers because $B_m = 0$ for all odd m other than 1. No such simple expression is known for odd positive integers.

The values of the zeta function obtained from integral arguments are called zeta constants. The following are the most commonly used values of the Riemann zeta function.

$$\zeta(0) = -\frac{1}{2},$$

$$\zeta(1/2) \approx -1.4603545 \quad \text{(sequence A059750 in OEIS)}$$

this is employed in calculating of kinetic boundary layer problems of linear kinetic equations.

$$\zeta(1) = 1 + \frac{1}{2} + \frac{1}{3} + \cdots = \infty;$$

if we approach from numbers larger 1. Then this is the harmonic series. But its principal value

$$\lim_{\varepsilon \to 0} \frac{\zeta(1+\varepsilon)+\zeta(1-\varepsilon)}{2}$$

exists which is the Euler-Mascheroni constant $\gamma = 0.5772....$

$\zeta(3/2) \approx 2.612$; (sequence A078434 in OEIS)

this is employed in calculating the critical temperature for a Bose–Einstein condensate in a box with periodic boundary conditions, and for spin wave physics in magnetic systems.

$\zeta(2) = 1 + \frac{1}{2^2} + \frac{1}{3^2} + \cdots = \frac{\pi^2}{6} \approx 1.645$; (sequence A013661 in OEIS)

the demonstration of this equality is known as the Basel problem. The reciprocal of this sum answers the question: What is the probability that two numbers selected at random are relatively prime?

$\zeta(3) = 1 + \frac{1}{2^3} + \frac{1}{3^3} + \cdots \approx 1.202$; (sequence A002117 in OEIS)

this is called Apéry's constant.

$\zeta(4) = 1 + \frac{1}{2^4} + \frac{1}{3^4} + \cdots = \frac{\pi^4}{90} \approx 1.0823$; (sequence A0013662 in OEIS)

This appears when integrating Planck's law to derive the Stefan–Boltzmann law in physics.

Euler Product Formula

The connection between the zeta function and prime numbers was discovered by Euler, who proved the identity

$$\sum_{n=1}^{\infty} \frac{1}{n^s} = \prod_{p \text{ prime}} \frac{1}{1-p^{-s}},$$

where, by definition, the left hand side is $\zeta(s)$ and the infinite product on the right hand side extends over all prime numbers p (such expressions are called Euler products):

$$\prod_{p \text{ prime}} \frac{1}{1-p^{-s}} = \frac{1}{1-2^{-s}} \cdot \frac{1}{1-3^{-s}} \cdot \frac{1}{1-5^{-s}} \cdot \frac{1}{1-7^{-s}} \cdot \frac{1}{1-11^{-s}} \cdots \frac{1}{1-p^{-s}} \cdots.$$

Both sides of the Euler product formula converge for $\mathrm{Re}(s) > 1$. The proof of Euler's identity uses only the formula for the geometric series and the fundamental theorem of arithmetic. Since the harmonic series, obtained when $s = 1$, diverges, Euler's formula (which becomes $\prod_p p/(p-1)$) implies that there are infinitely many primes.

The Euler product formula can be used to calculate the asymptotic probability that s randomly selected integers are set-wise coprime. Intuitively, the probability that any single number is divisible by a prime (or any integer), p is $1/p$. Hence the probability that s numbers are all divisible by this prime is $1/p^s$, and the probability that at least one of them is *not* is $1 - 1/p^s$. Now, for distinct primes, these divisibility events are mutually independent because the candidate divisors are coprime (a number is divisible by coprime divisors n and m if and only if it is divisible by nm, an event which occurs with probability $1/(nm)$.) Thus the asymptotic probability that s numbers are coprime is given by a product over all primes,

$$\prod_p^{\infty}\left(1-\frac{1}{p^s}\right)=\left(\prod_p^{\infty}\frac{1}{1-p^{-s}}\right)^{-1}=\frac{1}{\zeta(s)}.$$

(More work is required to derive this result formally.)

The Functional Equation

The Riemann zeta function satisfies the functional equation (also called the Riemann('s) functional equation)

$$\zeta(s) = 2^s \pi^{s-1} \sin\left(\frac{\pi s}{2}\right) \Gamma(1-s)\, \zeta(1-s),$$

where $\Gamma(s)$ is the gamma function, which is an equality of meromorphic functions valid on the whole complex plane. This equation relates values of the Riemann zeta function at the points s and $1 - s$. The functional equation (owing to the properties of the sine function) implies that $\zeta(s)$ has a simple zero at each even negative integer $s = -2n$ — these are known as the trivial zeros of $\zeta(s)$. For s an even positive integer, the product $\sin(\pi s/2)\Gamma(1-s)$ is regular and the functional equation relates the values of the Riemann zeta function at odd negative integers and even positive integers.

The functional equation was established by Riemann in his 1859 paper *On the Number of Primes Less Than a Given Magnitude* and used to construct the analytic continuation in the first place. An equivalent relationship had been conjectured by Euler over a hundred years earlier, in 1749, for the Dirichlet eta function (alternating zeta function)

$$\eta(s) = \sum_{n=1}^{\infty} \frac{(-1)^{n+1}}{n^s} = (1 - 2^{1-s})\zeta(s).$$

Incidentally, this relation is interesting also because it actually exhibits $\zeta(s)$ as a Dirichlet series (of the η-function) which is convergent (albeit non-absolutely) in the larger half-plane $\sigma > 0$ (not just $\sigma > 1$), up to an elementary factor.

Riemann also found a symmetric version of the functional equation, given by first defining

$$\xi(s) = \frac{1}{2}\pi^{-s/2} s(s-1)\Gamma\left(\frac{s}{2}\right)\zeta(s).$$

The functional equation is then given by

$$\xi(s) = \xi(1-s).$$

(Riemann defined a similar but different function which he called $\xi(t)$.)

Chapter 6

Riemann Hypothesis

In mathematics, the Riemann hypothesis, proposed by Bernhard Riemann (1859), is a conjecture that the nontrivial zeros of the Riemann zeta function all have real part 1/2. The name is also used for some closely related analogues, such as the Riemann hypothesis for curves over finite fields.

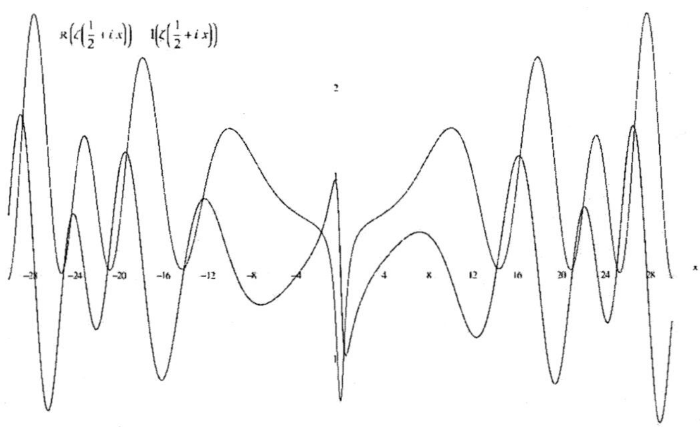

Figure: *The real part (red) and imaginary part (blue) of the Riemann zeta function along the critical line Re(s) = 1/2. The first non-trivial zeros can be seen at Im(s) = ±14.135, ±21.022 and ±25.011.*

The Riemann hypothesis implies results about the distribution of prime numbers. Along with suitable generalizations, it is considered by some mathematicians to be the most important unresolved problem in pure mathematics (Bombieri 2000). The Riemann hypothesis, along with the Goldbach conjecture, is part of Hilbert's eighth problem in David Hilbert's list of 23 unsolved problems; it is also one of the Clay Mathematics Institute Millennium Prize Problems.

The Riemann zeta function $\zeta(s)$ is defined for all complex numbers $s \neq 1$ with a simple pole at $s = 1$. It has zeros at the negative even integers (i.e. at $s = -2, -4, -6, \ldots$). These are called the trivial zeros. The Riemann hypothesis is concerned with the non-trivial zeros, and states that:

The real part of any non-trivial zero of the Riemann zeta function is 1/2.

Thus the non-trivial zeros should lie on the critical line, $1/2 + i\,t$, where t is a real number and i is the imaginary unit.

There are several nontechnical books on the Riemann hypothesis, such as Derbyshire (2003), Rockmore (2005), Sabbagh (2003), du Sautoy (2003). The books Edwards (1974), Patterson (1988) and Borwein et al. (2008) give mathematical introductions, while Titchmarsh (1986), Iviζ (1985) and Karatsuba & Voronin (1992) are advanced monographs.

The Riemann zeta function is defined for complex s with real part greater than 1 by the absolutely convergent infinite series

$$\zeta(s) = \sum_{n=1}^{\infty} \frac{1}{n^s} = \frac{1}{1^s} + \frac{1}{2^s} + \frac{1}{3^s} + \cdots.$$

Leonhard Euler showed that this series equals the Euler product

$$\zeta(s) = \prod_{p \text{ prime}} \frac{1}{1-p^{-s}} = \frac{1}{1-2^{-s}} \cdot \frac{1}{1-3^{-s}} \cdot \frac{1}{1-5^{-s}} \cdot \frac{1}{1-7^{-s}} \cdots \frac{1}{1-p^{-s}} \cdots$$

where the infinite product extends over all prime numbers p, and again converges for complex s with real part greater than 1. The convergence of the Euler product shows that $\zeta(s)$ has no zeros in this region, as none of the factors have zeros.

The Riemann hypothesis discusses zeros outside the region of convergence of this series, so it needs to be analytically continued to all complex s. This can be done by expressing it in terms of the Dirichlet eta function as follows. If s is greater than one, then the zeta function satisfies

$$\left(1-\frac{2}{2^s}\right)\zeta(s)=\sum_{n=1}^{\infty}\frac{(-1)^{n+1}}{n^s}=\frac{1}{1^s}-\frac{1}{2^s}+\frac{1}{3^s}-\cdots.$$

However, the series on the right converges not just when s is greater than one, but more generally whenever s has positive real part. Thus, this alternative series extends the zeta function from $\mathrm{Re}(s) > 1$ to the larger domain $\mathrm{Re}(s) > 0$, excluding the zeros $s = 1 + 2\pi in/\ln(2)$ of $1 - 2/2^s$.

In the strip $0 < \mathrm{Re}(s) < 1$ the zeta function also satisfies the functional equation

$$\zeta(s) = 2^s \pi^{s-1} \sin\left(\frac{\pi s}{2}\right) \Gamma(1-s)\, \zeta(1-s).$$

One may then define $\zeta(s)$ for all remaining nonzero complex numbers s by assuming that this equation holds outside the strip as well, and letting $\zeta(s)$ equal the right-hand side of the equation whenever s has non-positive real part. If s is a negative even integer then $\zeta(s) = 0$ because the factor $\sin(\pi s/2)$ vanishes; these are the trivial zeros of the zeta function. (If s is a positive even integer this argument does not apply because the zeros of sin are cancelled by the poles of the gamma function as it takes negative integer arguments.) The value $\zeta(0) = -1/2$ is not determined by the functional equation, but is the limiting value of $\zeta(s)$ as s approaches zero. The functional equation also implies that the zeta function has no zeros with negative real part other than the trivial zeros, so all non-trivial zeros lie in the critical strip where s has real part between 0 and 1.

History

> *"...es ist sehr wahrscheinlich, dass alle Wurzeln reell sind. Hiervon wäre allerdings ein strenger Beweis zu wünschen; ich habe indess die Aufsuchung desselben nach einigen flüchtigen vergeblichen Versuchen vorläufig bei Seite gelassen, da er für den nächsten Zweck meiner Untersuchung entbehrlich schien."*
>
> *"...it is very probable that all roots are real. Of course one would wish for a rigorous proof here; I have for the time being, after some fleeting vain attempts, provisionally put aside the search*

for this, as it appears dispensable for the next objective of my investigation."

Riemann's statement of the Riemann hypothesis, from (Riemann 1859). (He was discussing a version of the zeta function, modified so that its roots are real rather than on the critical line.)

In his 1859 paper *On the Number of Primes Less Than a Given Magnitude* Riemann found an explicit formula for the number of primes $\pi(x)$ less than a given number x. His formula was given in terms of the related function

$$\Pi(x) = \pi(x) + \frac{1}{2}\pi(x^{1/2}) + \frac{1}{3}\pi(x^{1/3})$$

$$+ \frac{1}{4}\pi(x^{1/4}) + \frac{1}{5}\pi(x^{1/5}) + \frac{1}{6}\pi(x^{1/6}) + \cdots$$

which counts primes where a prime power p^n counts as $1/n$ of a prime. The number of primes can be recovered from this function by

$$\pi(x) = \sum_{n=1}^{\infty} \frac{\mu(n)}{n}\Pi(x^{1/n}) = \Pi(x) - \frac{1}{2}\Pi(x^{1/2})$$

$$-\frac{1}{3}\Pi(x^{1/3}) - \frac{1}{5}\Pi(x^{1/5}) + \frac{1}{6}\Pi(x^{1/6}) - \cdots,$$

where μ is the Möbius function. Riemann's formula is then

$$\Pi_0(x) = \mathrm{Li}(x) - \sum_{\rho}\mathrm{Li}(x^{\rho}) - \log(2) + \int_x^{\infty} \frac{dt}{t(t^2-1)\log(t)}$$

where the sum is over the nontrivial zeros of the zeta function and where Π_0 is a slightly modified version of Π that replaces its value at its points of discontinuity by the average of its upper and lower limits:

$$\Pi_0(x) = \lim_{\varepsilon \to 0} \frac{\Pi(x-\varepsilon) + \Pi(x+\varepsilon)}{2}.$$

The summation in Riemann's formula is not absolutely convergent, but may be evaluated by taking the zeros ρ in order of the absolute value of their imaginary part. The function Li occurring in the first term is the (unoffset) logarithmic integral function given by the Cauchy principal value of the divergent integral

$$\mathrm{Li}(x) = \int_0^x \frac{dt}{\log(t)}.$$

The terms $\mathrm{Li}(x^\rho)$ involving the zeros of the zeta function need some care in their definition as Li has branch points at 0 and 1, and are defined (for $x > 1$) by analytic continuation in the complex variable ρ in the region Re(ρ) > 0, i.e. they should be considered as Ei(ρ ln x). The other terms also correspond to zeros: the dominant term Li(x) comes from the pole at $s = 1$, considered as a zero of multiplicity −1, and the remaining small terms come from the trivial zeros.

This formula says that the zeros of the Riemann zeta function control the oscillations of primes around their "expected" positions. Riemann knew that the non-trivial zeros of the zeta function were symmetrically distributed about the line $s = 1/2 + it$, and he knew that all of its non-trivial zeros must lie in the range $0 \leq \mathrm{Re}(s) \leq 1$. He checked that a few of the zeros lay on the critical line with real part 1/2 and suggested that they all do; this is the Riemann hypothesis.

Consequences of the Riemann Hypothesis

The practical uses of the Riemann hypothesis include many propositions which are known to be true under the Riemann hypothesis, and some which can be shown to be equivalent to the Riemann hypothesis.

Distribution of prime numbers

Riemann's explicit formula for the number of primes less than a given number in terms of a sum over the zeros of the Riemann zeta function says that the magnitude of the oscillations of primes around their expected position is controlled by the real parts of the zeros of the zeta function. In particular the error term in the prime number theorem is closely related to the position of the zeros: for example, the supremum of real parts of the zeros is the infimum of numbers β such that the error is $O(x^\beta)$ (Ingham 1932).

Von Koch (1901) proved that the Riemann hypothesis is equivalent to the "best possible" bound for the error of the prime number theorem.

A precise version of Koch's result, due to Schoenfeld (1976), says that the Riemann hypothesis is equivalent to

$$|\pi(x) - \mathrm{Li}(x)| < \frac{1}{8\pi}\sqrt{x}\log(x), \qquad \text{for all } x \geq 2657.$$

Growth of Arithmetic Functions

The Riemann hypothesis implies strong bounds on the growth of many other arithmetic functions, in addition to the primes counting function above.

One example involves the Möbius function μ. The statement that the equation

$$\frac{1}{\zeta(s)} = \sum_{n=1}^{\infty} \frac{\mu(n)}{n^s}$$

is valid for every s with real part greater than 1/2, with the sum on the right hand side converging, is equivalent to the Riemann hypothesis. From this we can also conclude that if the Mertens function is defined by

$$M(x) = \sum_{n \leq x} \mu(n)$$

then the claim that

$$M(x) = O(x^{1/2+\varepsilon})$$

for every positive ε is equivalent to the Riemann hypothesis (Titchmarsh 1986). The determinant of the order n Redheffer matrix is equal to $M(n)$, so the Riemann hypothesis can also be stated as a condition on the growth of these determinants. The Riemann hypothesis puts a rather tight bound on the growth of M, since Odlyzko & te Riele (1985) disproved the slightly stronger Mertens conjecture

$$|M(x)| \leq \sqrt{x}.$$

The Riemann hypothesis is equivalent to many other conjectures about the rate of growth of other arithmetic functions aside from $\mu(n)$. A typical example is Robin's theorem (Robin 1984), which states that if $\sigma(n)$ is the divisor function, given by

$$\sigma(n) = \sum_{d|n} d$$

then

$$\sigma(n) < e^{\gamma} n \log\log n$$

for all $n > 5040$ if and only if the Riemann hypothesis is true, where γ is the Euler–Mascheroni constant.

Another example was found by Jérôme Franel, and extended by Landau. The Riemann hypothesis is equivalent to several statements showing that the terms of the Farey sequence are fairly regular. One such equivalence is as follows: if F_n is the Farey sequence of order n, beginning with $1/n$ and up to $1/1$, then the claim that for all $\varepsilon > 0$

$$\sum_{i=1}^{m} |F_n(i) - i/m| = O(n^{1/2+\epsilon})$$

is equivalent to the Riemann hypothesis. Here $m = \sum_{i=1}^{n} \phi(i)$ is the number of terms in the Farey sequence of order n.

For an example from group theory, if $g(n)$ is Landau's function given by the maximal order of elements of the symmetric group S_n of degree n, then Massias, Nicolas & Robin (1988) showed that the Riemann hypothesis is equivalent to the bound

$$\log g(n) < \sqrt{\operatorname{Li}^{-1}(n)} \text{ for all sufficiently large } n.$$

Lindelöf Hypothesis and Growth of the Zeta Function

The Riemann hypothesis has various weaker consequences as well; one is the Lindelöf hypothesis on the rate of growth of the zeta function on the critical line, which says that, for any $\varepsilon > 0$,

$$\zeta\left(\frac{1}{2} + it\right) = O(t^{\varepsilon}),$$

as t tends to infinity.

The Riemann hypothesis also implies quite sharp bounds for the growth rate of the zeta function in other regions of the critical strip. For example, it implies that

$$e^{\gamma} \le \limsup_{t\to+\infty} \frac{|\zeta(1+it)|}{\log\log t} \le 2e^{\gamma}$$

$$\frac{6}{\pi^2} e^{\gamma} \le \limsup_{t\to+\infty} \frac{1/|\zeta(1+it)|}{\log\log t} \le \frac{12}{\pi^2} e^{\gamma}$$

so the growth rate of $\zeta(1+it)$ and its inverse would be known up to a factor of 2 (Titchmarsh 1986).

Large Prime Gap Conjecture

The prime number theorem implies that on average, the gap between the prime p and its successor is $\log p$. However, some gaps between primes may be much larger than the average. Cramér proved that, assuming the Riemann hypothesis, every gap is $O(\sqrt{p}\log p)$. This is a case in which even the best bound that can be proved using the Riemann Hypothesis is far weaker than what seems to be true: Cramér's conjecture implies that every gap is $O((\log p)^2)$ which, while larger than the average gap, is far smaller than the bound implied by the Riemann hypothesis. Numerical evidence supports Cramér's conjecture (Nicely 1999).

Criteria Equivalent to the Riemann Hypothesis

Many statements equivalent to the Riemann hypothesis have been found, though so far none of them have led to much progress in proving (or disproving) it. Some typical examples are as follows. (Others involve the divisor function $\sigma(n)$.)

The Riesz criterion was given by Riesz (1916), to the effect that the bound

$$-\sum_{k=1}^{\infty}\frac{(-x)^k}{(k-1)!\zeta(2k)} = O\left(x^{1/4+\epsilon}\right)$$

holds for all $\epsilon > 0$ if and only if the Riemann hypothesis holds.

Nyman (1950) proved that the Riemann Hypothesis is true if and only if the space of functions of the form

$$f(x) = \sum_{\nu=1}^{n} c_\nu \rho(\theta_\nu / x)$$

where $\rho(z)$ is the fractional part of z, $0 \le \theta_\nu \le 1$, and

$$\sum_{\nu=1}^{n} c_\nu \theta_\nu = 0,$$

is dense in the Hilbert space $L^2(0,1)$ of square-integrable functions on the unit interval. Beurling (1955) extended this by showing that the zeta function has no zeros with real part greater than $1/p$ if and only if this function space is dense in $L^p(0,1)$

Salem (1953) showed that the Riemann hypothesis is true if and only if the integral equation

$$\int_0^\infty \frac{z^{-\sigma-1}\phi(z)dz}{e^{x/z}+1} = 0$$

has no non-trivial bounded solutions φ for 1/2<σ<1.

Weil's criterion is the statement that the positivity of a certain function is equivalent to the Riemann hypothesis. Related is Li's criterion, a statement that the positivity of a certain sequence of numbers is equivalent to the Riemann hypothesis.

Speiser (1934) proved that the Riemann hypothesis is equivalent to the statement that $\zeta(s)$, the derivative of $\zeta'(s)$, has no zeros in the strip

$$0 < \Re(s) < \frac{1}{2}.$$

That ζ has only simple zeros on the critical line is equivalent (by definition) to its derivative having no zeros on the critical line.

Consequences of the Generalized Riemann Hypothesis

Several applications use the generalized Riemann hypothesis for Dirichlet L-series or zeta functions of number fields rather than just the Riemann hypothesis. Many basic properties of the Riemann zeta function can easily be generalized to all Dirichlet L-series, so it is plausible that a method that proves the Riemann hypothesis for the Riemann zeta function would also work for the generalized Riemann hypothesis for Dirichlet L-functions. Several results first proved using the generalized Riemann hypothesis were later given unconditional proofs without using it, though these were usually much harder. Many of the consequences on the following list are taken from Conrad (2010).

- In 1913, Gronwall showed that the generalized Riemann hypothesis implies that Gauss's list of imaginary quadratic fields with class number 1 is complete, though Baker, Stark and Heegner later gave unconditional proofs of this without using the generalized Riemann hypothesis.

- In 1917, Hardy and Littlewood showed that the generalized Riemann hypothesis implies a conjecture of Chebyshev that

  $$\lim_{x\to 1^-} \sum_{p>2} (-1)^{(p+1)/2} x^p = +\infty,$$

  which says that in some sense primes 3 mod 4 are more common than primes 1 mod 4.
- In 1923 Hardy and Littlewood showed that the generalized Riemann hypothesis implies a weak form of the Goldbach conjecture for odd numbers: that every sufficiently large odd number is the sum of three primes, though in 1937 Vinogradov gave an unconditional proof. In 1997 Deshouillers, Effinger, te Riele, and Zinoviev showed that the generalized Riemann hypothesis implies that every odd number greater than 5 is the sum of three primes.
- In 1934, Chowla showed that the generalized Riemann hypothesis implies that the first prime in the arithmetic progression a mod m is at most $Km^2\log(m)^2$ for some fixed constant K.
- In 1967, Hooley showed that the generalized Riemann hypothesis implies Artin's conjecture on primitive roots.
- In 1973, Weinberger showed that the generalized Riemann hypothesis implies that Euler's list of idoneal numbers is complete.
- Weinberger (1973) showed that the generalized Riemann hypothesis for the zeta functions of all algebraic number fields implies that any number field with class number 1 is either Euclidean or an imaginary quadratic number field of discriminant –19, –43, –67, or –163.
- In 1976, G. Miller showed that the generalized Riemann hypothesis implies that one can test if a number is prime in polynomial time via the Miller test. In 2002, Manindra Agrawal, Neeraj Kayal and Nitin Saxena proved this result unconditionally using the AKS primality test.

- Odlyzko (1990) discussed how the generalized Riemann hypothesis can be used to give sharper estimates for discriminants and class numbers of number fields.
- Ono & Soundararajan (1997) showed that the generalized Riemann hypothesis implies that Ramanujan's integral quadratic form $x^2 + y^2 + 10z^2$ represents all integers that it represents locally, with exactly 18 exceptions.

Excluded Middle

Some consequences of the RH are also consequences of its negation, and are thus theorems. In their discussion of the Hecke, Deuring, Mordell, Heilbronn theorem, (Ireland & Rosen 1990, p. 359) say

The method of proof here is truly amazing. If the generalized Riemann hypothesis is true, then the theorem is true. If the generalized Riemann hypothesis is false, then the theorem is true. Thus, the theorem is true!! (punctuation in original)

Care should be taken to understand what is meant by saying the generalized Riemann hypothesis is false: exactly which class of Dirichlet series is supposed to have a counterexample should be clearly indicated to avoid confusion.

Littlewood's Theorem

This concerns the sign of the error in the prime number theorem. It has been computed that $\pi(x) < \mathrm{Li}(x)$ for all $x \leq 10^{23}$, and no value of x is known for which $\pi(x) > \mathrm{Li}(x)$. See this table.

In 1914 Littlewood proved that there are arbitrarily large values of x for which

$$\pi(x) > \mathrm{Li}(x) + \frac{1}{3}\frac{\sqrt{x}}{\log x}\log\log\log x,$$

and that there are also arbitrarily large values of x for which

$$\pi(x) < \mathrm{Li}(x) - \frac{1}{3}\frac{\sqrt{x}}{\log x}\log\log\log x.$$

Thus the difference $\pi(x) - \mathrm{Li}(x)$ changes sign infinitely many times. Skewes' number is an estimate of the value of x

corresponding to the first sign change. His proof is divided into two cases: the RH is assumed to be false (about half a page of Ingham 1932, Chapt. V), and the RH is assumed to be true (about a dozen pages).

Gauss's Class Number Conjecture

This is the conjecture (first stated in article 303 of Gauss's Disquisitiones Arithmeticae) that there are only a finite number of imaginary quadratic fields with a given class number. One way to prove it would be to show that as the discriminant $D \to -\infty$ the class number $h(D) \to \infty$.

The following sequence of theorems involving the Riemann hypothesis is described in Ireland & Rosen 1990, pp. 358–361:

Hecke (1918)

Let $D < 0$ be the discriminant of an imaginary quadratic number field K. Assume the generalized Riemann hypothesis for L-functions of all imaginary quadratic Dirichlet characters. Then there is an absolute constant C such that

$$h(D) > C\frac{\sqrt{|D|}}{\log|D|}.$$

Deuring (1933)

If the RH is false then $h(D) > 1$ if $|D|$ is sufficiently large.

Mordell (1934)

If the RH is false then $h(D) \to \infty$ as $D \to -\infty$.

Heilbronn (1934)

If the generalized RH is false for the L-function of some imaginary quadratic Dirichlet character then $h(D) \to \infty$ as $D \to -\infty$.

(In the work of Hecke and Heilbronn, the only L-functions that occur are those attached to imaginary quadratic characters, and it is only for those L-functions that *GRH is true* or *GRH is false* is intended; a failure of GRH for the L-function of a cubic Dirichlet character would, strictly speaking, mean GRH is false, but that was not the kind of failure of GRH that Heilbronn had in mind, so his assumption was more restricted than simply *GRH is false*.)

In 1935, Carl Siegel later strengthened the result without using RH or GRH in any way.

Growth of Euler's Totient

In 1983 J. L. Nicolas proved (Ribenboim 1996, p. 320) that

$$\phi(n) < e^{-\gamma} \frac{n}{\log \log n} \quad \text{for infinitely many } n,$$

where $\varphi(n)$ is Euler's totient function and γ is Euler's constant.

Ribenboim Remarks That

The method of proof is interesting, in that the inequality is shown first under the assumption that the Riemann hypothesis is true, secondly under the contrary assumption.

Generalizations and Analogs of the Riemann Hypothesis

Dirichlet L-series and Other Number Fields: The Riemann hypothesis can be generalized by replacing the Riemann zeta function by the formally similar, but much more general, global L-functions. In this broader setting, one expects the non-trivial zeros of the global L-functions to have real part 1/2. It is these conjectures, rather than the classical Riemann hypothesis only for the single Riemann zeta function, which accounts for the true importance of the Riemann hypothesis in mathematics.

The generalized Riemann hypothesis extends the Riemann hypothesis to all Dirichlet L-functions. In particular it implies the conjecture that Siegel zeros (zeros of L functions between 1/2 and 1) do not exist.

The extended Riemann hypothesis extends the Riemann hypothesis to all Dedekind zeta functions of algebraic number fields. The extended Riemann hypothesis for abelian extension of the rationals is equivalent to the generalized Riemann hypothesis. The Riemann hypothesis can also be extended to the L-functions of Hecke characters of number fields.

The grand Riemann hypothesis extends it to all automorphic zeta functions, such as Mellin transforms of Hecke eigenforms.

Function Fields and Zeta Functions of Varieties over Finite Fields

Artin (1924) introduced global zeta functions of (quadratic) function fields and conjectured an analogue of the Riemann hypothesis for them, which has been proven by Hasse in the genus 1 case and by Weil (1948) in general. For instance, the fact that the Gauss sum, of the quadratic character of a finite field of size q (with q odd), has absolute value

$$\sqrt{q}$$

is actually an instance of the Riemann hypothesis in the function field setting. This led Weil (1949) to conjecture a similar statement for all algebraic varieties; the resulting Weil conjectures were proven by Pierre Deligne (1974, 1980).

Arithmetic Zeta Functions of Arithmetic Schemes and their L-factors

Arithmetic zeta functions generalise the Riemann and Dedekind zeta functions as well as the zeta functions of varieties over finite fields to every arithmetic scheme or a scheme of finite type over integers. The arithmetic zeta function of a regular connected equidimensional arithmetic scheme of Kronecker dimension $n$can be factorized into the product of appropriately defined L-factors and an auxiliary factor Jean-Pierre Serre (1970). Assuming a functional equation and meromorphic continuation, the generalized Riemann hypothesis for the L-factor states that its zeros inside the critical strip $\Re(s) \in (0,n)$ lie on the central line. Correspondingly, the generalized Riemann hypothesis for the arithmetic zeta function of a regular connected equidimensional arithmetic scheme states that its zeros inside the critical strip lie on vertical lines $\Re(s) = 1/2, 3/2, \ldots, n-1/2$ and its poles inside the critical strip lie on vertical lines $\Re(s) = 1, 2, \ldots, n-1$. This is known for schemes in positive characteristic and follows from Pierre Deligne (1974, 1980), but remains entirely unknown in characteristic zero.

Selberg Zeta Functions

Selberg (1956) introduced the Selberg zeta function of a Riemann surface. These are similar to the Riemann zeta function: they have a functional equation, and an infinite product similar to the Euler product but taken over closed geodesics rather than primes.

The Selberg trace formula is the analogue for these functions of the explicit formulas in prime number theory. Selberg proved that the Selberg zeta functions satisfy the analogue of the Riemann hypothesis, with the imaginary parts of their zeros related to the eigenvalues of the Laplacian operator of the Riemann surface.

Ihara Zeta Functions

The Ihara zeta function of a finite graph is an analogue of the Selberg zeta function introduced by Yasutaka Ihara. A regular finite graph is a Ramanujan graph, a mathematical model of efficient communication networks, if and only if its Ihara zeta function satisfies the analogue of the Riemann hypothesis as was pointed out by T. Sunada.

Montgomery's Pair Correlation Conjecture

Montgomery (1973) suggested the pair correlation conjecture that the correlation functions of the (suitably normalized) zeros of the zeta function should be the same as those of the eigenvalues of a random hermitian matrix. Odlyzko (1987) showed that this is supported by large scale numerical calculations of these correlation functions.

Montgomery showed that (assuming the Riemann hypothesis) at least 2/3 of all zeros are simple, and a related conjecture is that all zeros of the zeta function are simple (or more generally have no non-trivial integer linear relations between their imaginary parts). Dedekind zeta functions of algebraic number fields, which generalize the Riemann zeta function, often do have multiple complex zeros. This is because the Dedekind zeta functions factorize as a product of powers of Artin L-functions, so zeros of Artin L-functions sometimes give rise to multiple zeros of Dedekind zeta functions. Other examples of zeta functions with multiple zeros are the L-functions of some elliptic curves: these can have multiple zeros at the real point of their critical line; the Birch-Swinnerton-Dyer conjecture predicts that the multiplicity of this zero is the rank of the elliptic curve.

Other Zeta Functions

There are many other examples of zeta functions with analogues of the Riemann hypothesis, some of which have been proved. Goss zeta functions of function fields have a Riemann hypothesis, proved by Sheats (1998). The main conjecture of Iwasawa theory, proved by Barry Mazur and Andrew Wiles for cyclotomic fields, and Wiles for totally real fields, identifies the zeros of a p-adic L-function with the eigenvalues of an operator, so can be thought of as an analogue of the Hilbert–Pólya conjecture for p-adic L-functions (Wiles 2000).

Attempts to Prove the Riemann Hypothesis

Several mathematicians have addressed the Riemann hypothesis, but none of their attempts have yet been accepted as correct solutions. Watkins (2007) lists some incorrect solutions, and more are frequently announced.

Operator Theory

Hilbert and Polya suggested that one way to derive the Riemann hypothesis would be to find a self-adjoint operator, from the existence of which the statement on the real parts of the zeros of $\zeta(s)$ would follow when one applies the criterion on real eigenvalues. Some support for this idea comes from several analogues of the Riemann zeta functions whose zeros correspond to eigenvalues of some operator: the zeros of a zeta function of a variety over a finite field correspond to eigenvalues of a Frobenius element on an étale cohomology group, the zeros of a Selberg zeta function are eigenvalues of a Laplacian operator of a Riemann surface, and the zeros of a p-adic zeta function correspond to eigenvectors of a Galois action on ideal class groups.

Odlyzko (1987) showed that the distribution of the zeros of the Riemann zeta function shares some statistical properties with the eigenvalues of random matrices drawn from the Gaussian unitary ensemble. This gives some support to the Hilbert–Pólya conjecture.

In 1999, Michael Berry and Jon Keating conjectured that there is some unknown quantization $\hat{H}$ of the classical Hamiltonian $H = xp$ so that

$$\zeta(1/2 + i\hat{H}) = 0$$

and even more strongly, that the Riemann zeros coincide with the spectrum of the operator $1/2 + i\hat{H}$. This is to be contrasted to canonical quantization which leads to the Heisenberg uncertainty principle $[x, p] = 1/2$ and the natural numbers as spectrum of the quantum harmonic oscillator. The crucial point is that the Hamiltonian should be a self-adjoint operator so that the quantization would be a realization of the Hilbert–Pólya programme. In a connection with this quantum mechanical problem Berry and Connes had proposed that the inverse of the potential of the Hamiltonian is connected to the half-derivative of the function $N(s) = \frac{1}{\pi}\operatorname{Arg}\xi(1/2 + i\sqrt{s})$ then, in Berry–Connes approach $V^{-1}(x) = \sqrt{4\pi}\frac{d^{1/2}N(x)}{dx^{1/2}}$ (Connes 1999). This yields to a Hamiltonian whose eigenvalues are the square of the imaginary part of the Riemann zeros, and also the functional determinant of this Hamiltonian operator is just the Riemann Xi function. In fact the Riemann Xi function would be proportional to the functional determinant (Hadamard product) $\det(H + 1/4 + s(s-1))$ as proven by Connes and others, in this approach $\frac{\xi(s)}{\xi(0)} = \frac{\det(H + s(s-1) + 1/4)}{\det(H + 1/4)}$.

The analogy with the Riemann hypothesis over finite fields suggests that the Hilbert space containing eigenvectors corresponding to the zeros might be some sort of first cohomology group of the spectrum Spec(Z) of the integers. Deninger (1998) described some of the attempts to find such a cohomology theory (Leichtnam 2005).

Zagier (1983) constructed a natural space of invariant functions on the upper half plane which has eigenvalues under the Laplacian operator corresponding to zeros of the Riemann zeta function, and remarked that in the unlikely event that one could show the existence of a suitable positive definite inner product on this space the Riemann hypothesis would follow. Cartier (1982) discussed a related example, where due to a bizarre bug a computer programme listed zeros of the Riemann zeta function as eigenvalues of the same Laplacian operator.

Schumayer & Hutchinson (2011) surveyed some of the attempts to construct a suitable physical model related to the Riemann zeta function.

Lee–Yang Theorem

The Lee–Yang theorem states that the zeros of certain partition functions in statistical mechanics all lie on a "critical line" with real part 0, and this has led to some speculation about a relationship with the Riemann hypothesis (Knauf 1999).

Turán's Result

Pál Turán (1948) showed that if the functions

$$\sum_{n=1}^{N} n^{-s}$$

have no zeros when the real part of s is greater than one then

$$T(x) = \sum_{n \le x} \frac{\lambda(n)}{n} \ge 0 \text{ for } x > 0,$$

where $\lambda(n)$ is the Liouville function given by $(-1)^r$ if n has r prime factors. He showed that this in turn would imply that the Riemann hypothesis is true. However Haselgrove (1958) proved that $T(x)$ is negative for infinitely many x (and also disproved the closely related Pólya conjecture), and Borwein, Ferguson & Mossinghoff (2008) showed that the smallest such x is 72185376951205. Spira (1968) showed by numerical calculation that the finite Dirichlet series above for N=19 has a zero with real part greater than 1. Turán also showed that a somewhat weaker assumption, the nonexistence of zeros with real part greater than $1+N^{-1/2+\varepsilon}$ for large N in the finite Dirichlet series above, would also imply the Riemann hypothesis, but Montgomery (1983) showed that for all sufficiently large N these series have zeros with real part greater than $1 + (\log \log N)/(4 \log N)$. Therefore, Turán's result is vacuously true and cannot be used to help prove the Riemann hypothesis.

Noncommutative Geometry

Connes (1999, 2000) has described a relationship between the Riemann hypothesis and noncommutative geometry, and shows that a suitable analog of the Selberg trace formula for the action of the idèle class group on the adèle class space would imply the Riemann hypothesis. Some of these ideas are elaborated in Lapidus (2008).

Hilbert Spaces of Entire Functions

Louis de Branges (1992) showed that the Riemann hypothesis would follow from a positivity condition on a certain Hilbert space of entire functions. However Conrey & Li (2000) showed that the necessary positivity conditions are not satisfied.

Quasicrystals

The Riemann hypothesis implies that the zeros of the zeta function form a quasicrystal, meaning a distribution with discrete support whose Fourier transform also has discrete support. Dyson (2009) suggested trying to prove the Riemann hypothesis by classifying, or at least studying, 1-dimensional quasicrystals.

Arithmetic Zeta Functions of Models of Elliptic Curves Over Number Fields

When one goes from geometric dimension one, e.g. an algebraic number field, to geometric dimension two, e.g. a regular model of an elliptic curve over a number field, the two-dimensional part of the generalized Riemann hypothesis for the arithmetic zeta function of the model deals with the poles of the zeta function. In dimension one the study of the zeta integral in Tate's thesis does not lead to new important information on the Riemann hypothesis. Contrary to this, in dimension two work of Ivan Fesenko on two-dimensional generalisation of Tate's thesis includes an integral representation of a zeta integral closely related to the zeta function. In this new situation, not possible in dimension one, the poles of the zeta function can be studied via the zeta integral and associated adele groups. Related conjecture of Fesenko (2010) on the positivity of the fourth derivative of a boundary function associated to the zeta integral essentially implies the pole part of the generalized Riemann hypothesis. Suzuki (2011) proved that the latter, together with some technical assumptions, implies Fesenko's conjecture.

Multiple Zeta Functions

Deligne's proof of the Riemann hypothesis over finite fields used the zeta functions of product varieties, whose zeros and poles correspond to sums of zeros and poles of the original zeta function, in order to bound the real parts of the zeros of the original zeta function. By analogy, Kurokawa (1992) introduced multiple zeta functions whose zeros and poles correspond to sums of zeros and poles of the Riemann zeta function. To make the series converge he restricted to sums of zeros or poles all with non-negative imaginary part. So far, the known bounds on the zeros and poles of the multiple zeta functions are not strong enough to give useful estimates for the zeros of the Riemann zeta function.

Location of the Zeros

Number of Zeros: The functional equation combined with the argument principle implies that the number of zeros of the zeta function with imaginary part between 0 and T is given by

$$N(T) = \frac{1}{\pi}\operatorname{Arg}(\xi(s)) = \frac{1}{\pi}\operatorname{Arg}(\Gamma(s/2)\pi^{-s/2}\zeta(s)s(s-1)/2)$$

for $s=1/2+iT$, where the argument is defined by varying it continuously along the line with $\mathrm{Im}(s)=T$, starting with argument 0 at $\infty+iT$. This is the sum of a large but well understood term

$$\frac{1}{\pi}\operatorname{Arg}(\Gamma(s/2)\pi^{-s/2}s(s-1)/2) = \frac{T}{2\pi}\log\frac{T}{2\pi} - \frac{T}{2\pi} + 7/8 + O(1/T)$$

and a small but rather mysterious term

$$S(T) = \frac{1}{\pi}\operatorname{Arg}(\zeta(1/2+iT)) = O(\log(T)).$$

So the density of zeros with imaginary part near T is about $\log(T)/2\pi$, and the function S describes the small deviations from this. The function $S(t)$ jumps by 1 at each zero of the zeta function, and for $t \geq 8$ it decreases monotonically between zeros with derivative close to "log t.

Karatsuba (1996) proved that every interval $(T, T+H]$ for $H \geq T^{27/82+\varepsilon}$ contains at least

$$H(\ln T)^{1/3} e^{-c\sqrt{\ln\ln T}}$$

points where the function $S(t)$ changes sign.

Selberg (1946) showed that the average moments of even powers of S are given by

$$\int_0^T |S(t)|^{2k}\,dt = \frac{(2k)!}{k!(2\pi)^{2k}} T(\log\log T)^k + O(T(\log\log T)^{k-1/2}).$$

This suggests that $S(T)/(\log\log T)^{1/2}$ resembles a Gaussian random variable with mean 0 and variance $2\pi^2$ (Ghosh (1983) proved this fact). In particular $|S(T)|$ is usually somewhere around $(\log\log T)^{1/2}$, but occasionally much larger. The exact order of growth of $S(T)$ is not known. There has been no unconditional improvement to Riemann's original bound $S(T)=O(\log T)$, though the Riemann hypothesis implies the slightly smaller bound $S(T)=O(\log T/\log\log T)$ (Titchmarsh 1985). The true order of magnitude may be somewhat less than this, as random functions with the same distribution as $S(T)$ tend to have growth of order about $\log(T)^{1/2}$. In the other direction it cannot be too small: Selberg (1946) showed that $S(T) \neq o((\log T)^{1/3}/(\log\log T)^{7/3})$, and assuming the Riemann hypothesis Montgomery showed that $S(T) \neq o((\log T)^{1/2}/(\log\log T)^{1/2})$.

Numerical calculations confirm that S grows very slowly: $|S(T)| < 1$ for $T < 280$, $|S(T)| < 2$ for $T < 6800000$, and the largest value of $|S(T)|$ found so far is not much larger than 3 (Odlyzko 2002).

Riemann's estimate $S(T) = O(\log T)$ implies that the gaps between zeros are bounded, and Littlewood improved this slightly, showing that the gaps between their imaginary parts tends to 0.

Theorem of Hadamard and de la Vallée-Poussin

Hadamard (1896) and de la Vallée-Poussin (1896) independently proved that no zeros could lie on the line $\mathrm{Re}(s) = 1$. Together with the functional equation and the fact that there are no zeros with real part greater than 1, this showed that all non-trivial zeros must lie in the interior of the critical strip $0 < \mathrm{Re}(s) < 1$. This was a key step in their first proofs of the prime number theorem.

Both the original proofs that the zeta function has no zeros with real part 1 are similar, and depend on showing that if $\zeta(1+it)$ vanishes, then $\zeta(1+2it)$ is singular, which is not possible. One way of doing this is by using the inequality

$|\zeta(\sigma)^3\zeta(\sigma+it)^4\zeta(\sigma+2it)| \geq 1$ for $\sigma>1$, t real, and looking at the limit as σ tends to 1. This inequality follows by taking the real part of the log of the Euler product to see that

$$|\zeta(\sigma+it)| = \exp \Re \sum_{p^n} \frac{p^{-n(\sigma+it)}}{n} = \exp \sum_{p^n} \frac{p^{-n\sigma}\cos(t\log p^n)}{n}$$

(where the sum is over all prime powers p^n) so that

$$|\zeta(\sigma)^3 \zeta(\sigma+it)^4 \zeta(\sigma+2it)| = \exp \sum_{p^n} p^{-n\sigma} \frac{3+4\cos(t\log p^n)+\cos(2t\log p^n)}{n}$$

which is at least 1 because all the terms in the sum are positive, due to the inequality

$$3+4\cos(\theta)+\cos(2\theta) = 2(1+\cos(\theta))^2 \geq 0.$$

Zero-free Regions

De la Vallée-Poussin (1899-1900) proved that if $\sigma+it$ is a zero of the Riemann zeta function, then $1-\sigma \geq C/\log(t)$ for some positive constant C. In other words zeros cannot be too close to the line $\sigma=1$: there is a zero-free region close to this line. This zero-free region has been enlarged by several authors. Ford (2002) gave a version with explicit numerical constants: $\zeta(\sigma + it) \neq 0$ whenever $|t| \geq 3$ and

$$\sigma \geq 1 - \frac{1}{57.54(\log|t|)^{2/3}(\log\log|t|)^{1/3}}.$$

Zeros on the Critical Line

Hardy (1914) and Hardy & Littlewood (1921) showed there are infinitely many zeros on the critical line, by considering moments of certain functions related to the zeta function. Selberg (1942) proved that at least a (small) positive proportion of zeros lie on the line. Levinson (1974) improved this to one-third of the zeros by relating the zeros of the zeta function to those of its derivative, and Conrey (1989) improved this further to two-fifths.

Most zeros lie close to the critical line. More precisely, Bohr & Landau (1914) showed that for any positive ε, all but an infinitely small proportion of zeros lie within a distance ε of the critical line. Iviζ (1985) gives several more precise versions of this result, called zero density estimates, which bound the number of zeros in regions with imaginary part at most T and real part at least $1/2+\varepsilon$.

Numerical Calculations

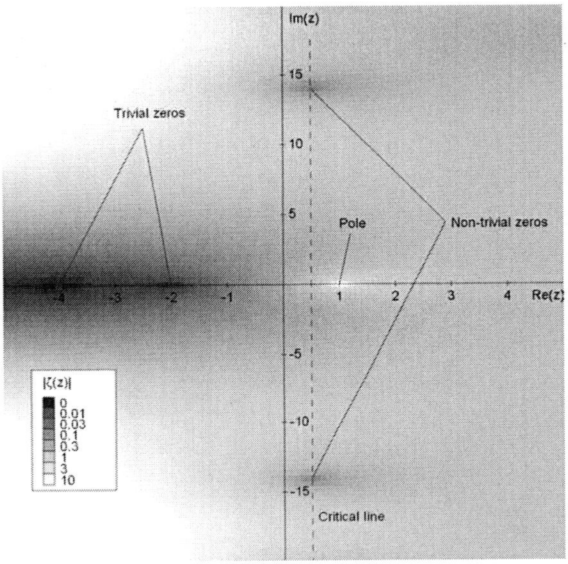

***Figure:** Absolute value of the ζ-function*

The function

$$\pi^{-s/2}\Gamma(s/2)\zeta(s)$$

has the same zeros as the zeta function in the critical strip, and is real on the critical line because of the functional equation, so one can prove the existence of zeros exactly on the real line between two points by checking numerically that the function has opposite signs at these points. Usually one writes

$$\zeta(1/2+it) = Z(t)e^{-i\pi\theta(t)}$$

where Hardy's function Z and the Riemann–Siegel theta function θ are uniquely defined by this and the condition that they are smooth real functions with $\theta(0)=0$. By finding many intervals where the function Z changes sign one can show that there are many zeros on the critical line. To verify the Riemann hypothesis up to a given imaginary part T of the zeros, one also has to check that there are no further zeros off the line in this region. This can be done by calculating the total number of zeros in the region and checking that it is the same as the number of zeros found on the line. This allows one to verify the Riemann hypothesis computationally up to any desired value of T (provided all the zeros of the zeta function in this region are simple and on the critical line).

Some calculations of zeros of the zeta function are listed below. So far all zeros that have been checked are on the critical line and are simple. (A multiple zero would cause problems for the zero finding algorithms, which depend on finding sign changes between zeros.)

Year	***Number of zeros***	***Author***
1859?	3	B. Riemann used the Riemann–Siegel formula (unpublished, but reported in Siegel 1932).
1903	15	J. P. Gram (1903) used Euler–Maclaurin summation and discovered Gram's law. He showed that all 10 zeros with imaginary part at most 50 range lie on the critical line with real part 1/2 by computing the sum of the inverse 10th powers of the roots he found.
1914	79 ($G_n \leq 200$)	R. J. Backlund (1914) introduced a better method of checking all the zeros up to that point are on the line, by studying the argument $S(T)$ of the zeta function.
1925	138 ($\gamma_n \leq 300$)	J. I. Hutchinson (1925) found the first failure of Gram's law, at the Gram point g_{126}.
1935	195	E. C. Titchmarsh (1935) used the recently rediscovered Riemann–Siegel formula, which is much faster than Euler–Maclaurin summation. It takes about $O(T^{3/2+e})$ steps to check zeros with imaginary part less than T, while the Euler–Maclaurin method takes about $O(T^{2+e})$ steps.
1936	1041	E. C. Titchmarsh (1936) and L. J. Comrie were the last to find zeros by hand.
1953	1104	A. M. Turing (1953) found a more efficient way to check that all zeros up to some point are accounted for by the zeros on the line, by checking that Z has the correct sign at several consecutive Gram points and using the fact that $S(T)$ has average value 0. This requires almost no extra work because the sign of Z at Gram points is already known from finding the zeros, and is still the usual method used. This was the first use of a digital computer to calculate the zeros.

Contd...

Year	***Number of zeros***	***Author***
1956	15000	D. H. Lehmer (1956) discovered a few cases where the zeta function has zeros that are "only just" on the line: two zeros of the zeta function are so close together that it is unusually difficult to find a sign change between them. This is called "Lehmer's phenomenon", and first occurs at the zeros with imaginary parts 7005.063 and 7005.101, which differ by only .04 while the average gap between other zeros near this point is about 1.
1956	25000	D. H. Lehmer
1958	35337	N. A. Meller
1966	250000	R. S. Lehman
1968	3500000	Rosser, Yohe & Schoenfeld (1969) stated Rosser's rule (described below).
1977	40000000	R. P. Brent
1979	81000001	R. P. Brent
1982	200000001	R. P. Brent, J. van de Lune, H. J. J. te Riele, D. T. Winter
1983	300000001	J. van de Lune, H. J. J. te Riele
1986	1500000001	van de Lune, te Riele & Winter (1986) gave some statistical data about the zeros and give several graphs of Z at places where it has unusual behaviour.
1987	A few of large ($\sim 10^{12}$) height	A. M. Odlyzko (1987) computed smaller numbers of zeros of much larger height, around 10^{12}, to high precision to check Montgomery's pair correlation conjecture.
1992	A few of large ($\sim 10^{20}$) height	A. M. Odlyzko (1992) computed a 175 million zeroes of heights around 10^{20} and a few more of heights around 2×10^{20}, and gave an extensive discussion of the results.
1998	10000 of large ($\sim 10^{21}$) height	A. M. Odlyzko (1998) computed some zeros of height about 10^{21}
2001	10000000000	J. van de Lune (unpublished)
2004	900000000000	S. Wedeniwski (ZetaGrid distributed computing)
2004	10000000000000 and a few of large (up to $\sim 10^{24}$) heights	X. Gourdon (2004) and Patrick Demichel used the Odlyzko–Schönhage algorithm. They also checked two billion zeros around heights 10^{13}, 10^{14}, ..., 10^{24}.

Gram Points

A Gram point is a point on the critical line $1/2 + it$ where the zeta function is real and non-zero. Using the expression for the zeta function on the critical line, $\xi(1/2 + it) = Z(t)e^{-i\theta(t)}$, where Hardy's function, Z, is real for real t, and θ is the Riemann–Siegel theta function, we see that zeta is real when $\sin(\theta(t)) = 0$. This implies that $\theta(t)$ is an integer multiple of p which allows for the location of Gram points to be calculated fairly easy by inverting the formula for θ. They are usually numbered as g_n for $n = 0, 1, ...$, where g_n is the unique solution of $\theta(t) = n\pi$.

Gram observed that there was often exactly one zero of the zeta function between any two Gram points; Hutchinson called this observation Gram's law. There are several other closely related statements that are also sometimes called Gram's law: for example, $(-1)^n Z(g_n)$ is usually positive, or $Z(t)$ usually has opposite sign at consecutive Gram points. The imaginary parts γ_n of the first few zeros (in blue) and the first few Gram points g_n are given in the following table

		$g_{?1}$	γ_1	g_0	γ_2	g_1	γ_3	g_2	γ_4	g_3	γ_5	g_4	γ_6	g_5
0.0 00	3.4 36	9.6 67	14.1 35	17.8 46	21.0 22	23.1 70	25.0 11	27.6 70	30.4 25	31.7 18	32.9 35	35.4 67	37.5 86	38.9 99

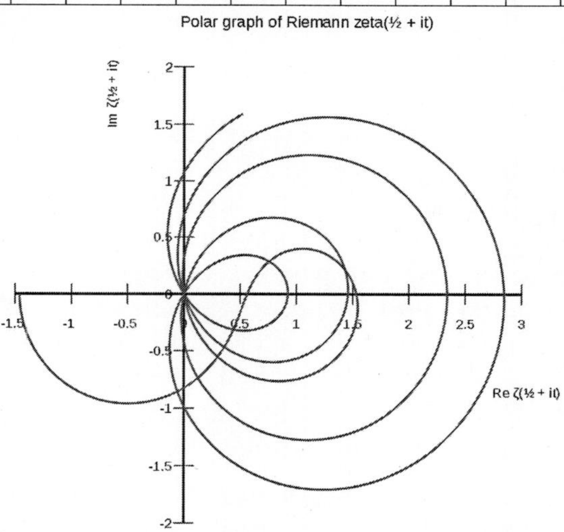

Figure: *This shows the values of $\zeta(1/2+it)$ in the complex plane for $0 \leq t \leq 34$. (For $t=0$, $\zeta(1/2) \approx -1.460$ corresponds to the leftmost point of the red curve.) Gram's law states that the curve usually crosses the real axis once between zeros.*

The first failure of Gram's law occurs at the 127'th zero and the Gram point g_{126}, which are in the "wrong" order.

g_{124}	γ_{126}	g_{125}	g_{126}	γ_{127}	γ_{128}	g_{127}	γ_{129}	g_{128}
279.148	279.229	280.802	282.455	282.465	283.211	284.104	284.836	285.752

A Gram point t is called good if the zeta function is positive at $1/2 + it$. The indices of the "bad" Gram points where Z has the "wrong" sign are 126, 134, 195, 211,... (sequence A114856 in OEIS). A Gram block is an interval bounded by two good Gram points such that all the Gram points between them are bad. A refinement of Gram's law called Rosser's rule due to Rosser, Yohe & Schoenfeld (1969) says that Gram blocks often have the expected number of zeros in them (the same as the number of Gram intervals), even though some of the individual Gram intervals in the block may not have exactly one zero in them. For example, the interval bounded by g_{125} and g_{127} is a Gram block containing a unique bad Gram point g_{126}, and contains the expected number 2 of zeros although neither of its two Gram intervals contains a unique zero. Rosser et al. checked that there were no exceptions to Rosser's rule in the first 3 million zeros, although there are infinitely many exceptions to Rosser's rule over the entire zeta function.

Gram's rule and Rosser's rule both say that in some sense zeros do not stray too far from their expected positions. The distance of a zero from its expected position is controlled by the function S defined above, which grows extremely slowly: its average value is of the order of $(\log \log T)^{1/2}$, which only reaches 2 for T around 10^{24}. This means that both rules hold most of the time for small T but eventually break down often. Indeed Trudgian (2011) showed that both Gram's law and Rosser's rule fail in a positive proportion of cases. To be more specific, it is expected that in about 73% one zero is enclosed by two successive Gram points, but in 14% no zero and in 13% two zeros are in such a Gram-interval on the long run.

Arguments for and Against the Riemann Hypothesis

Mathematical papers about the Riemann hypothesis tend to be cautiously noncommittal about its truth. Of authors who express an opinion, most of them, such as Riemann (1859) or Bombieri (2000), imply that they expect (or at least hope) that it is true. The few authors who express serious doubt about it include Iviæ (2008) who lists some reasons for being skeptical, and Littlewood (1962) who flatly states that he believes it to be false, and that there is no evidence whatever for it and no imaginable reason for it to be true. The consensus of the survey articles (Bombieri 2000, Conrey 2003, and Sarnak 2008) is that the evidence for it is strong but not overwhelming, so that while it is probably true there is some reasonable doubt about it.

Some of the arguments for (or against) the Riemann hypothesis are listed by Sarnak (2008), Conrey (2003), and Iviæ (2008), and include the following reasons.

- Several analogues of the Riemann hypothesis have already been proved. The proof of the Riemann hypothesis for varieties over finite fields by Deligne (1974) is possibly the single strongest theoretical reason in favor of the Riemann hypothesis. This provides some evidence for the more general conjecture that all zeta functions associated with automorphic forms satisfy a Riemann hypothesis, which includes the classical Riemann hypothesis as a special case. Similarly Selberg zeta functions satisfy the analogue of the Riemann hypothesis, and are in some ways similar to the Riemann zeta function, having a functional equation and an infinite product expansion analogous to the Euler product expansion. However there are also some major differences; for example they are not given by Dirichlet series. The Riemann hypothesis for the Goss zeta function was proved by Sheats (1998). In contrast to these positive examples, however, some Epstein zeta functions do not satisfy the Riemann hypothesis, even though they have an infinite number of zeros on the critical line (Titchmarsh 1986). These functions are quite similar to the Riemann zeta function, and have a Dirichlet series expansion and a functional equation, but the ones known to fail the Riemann hypothesis do not have an Euler product and are not directly related to automorphic representations.
- The numerical verification that many zeros lie on the line seems at first sight to be strong evidence for it. However analytic number theory has had many

conjectures supported by large amounts of numerical evidence that turn out to be false. See Skewes number for a notorious example, where the first exception to a plausible conjecture related to the Riemann hypothesis probably occurs around 10^{316}; a counterexample to the Riemann hypothesis with imaginary part this size would be far beyond anything that can currently be computed. The problem is that the behaviour is often influenced by very slowly increasing functions such as log log T, that tend to infinity, but do so so slowly that this cannot be detected by computation. Such functions occur in the theory of the zeta function controlling the behaviour of its zeros; for example the function $S(T)$ above has average size around $(\log \log T)^{1/2}$. As $S(T)$ jumps by at least 2 at any counterexample to the Riemann hypothesis, one might expect any counterexamples to the Riemann hypothesis to start appearing only when $S(T)$ becomes large. It is never much more than 3 as far as it has been calculated, but is known to be unbounded, suggesting that calculations may not have yet reached the region of typical behaviour of the zeta function.

- Denjoy's probabilistic argument for the Riemann hypothesis (Edwards 1974) is based on the observation that if $\mu(x)$ is a random sequence of "1"s and "–1"s then, for every $\varepsilon > 0$, the partial sums

  $$M(x) = \sum_{n \le x} \mu(n)$$

  (the values of which are positions in a simple random walk) satisfy the bound

  $$M(x) = O(x^{1/2+\varepsilon})$$

  with probability 1. The Riemann hypothesis is equivalent to this bound for the Möbius function μ and the Mertens function M derived in the same way from it. In other words, the Riemann hypothesis is in some sense equivalent to saying that $\mu(x)$ behaves like a random sequence of coin tosses. When $\mu(x)$ is non-zero its sign gives the parity of the number of prime factors of x, so informally the Riemann hypothesis says that the parity of the number of prime factors of an integer behaves randomly. Such probabilistic arguments in number theory often give the right answer, but tend to be very hard to make rigorous, and occasionally give the wrong answer for some results, such as Maier's theorem.

- The calculations in Odlyzko (1987) show that the zeros of the zeta function behave very much like the eigenvalues of a random Hermitian matrix, suggesting that they are the eigenvalues of some self-adjoint operator, which would imply the Riemann hypothesis. However all attempts to find such an operator have failed.
- There are several theorems, such as Goldbach's conjecture for sufficiently large odd numbers, that were first proved using the generalized Riemann hypothesis, and later shown to be true unconditionally. This could be considered as weak evidence for the generalized Riemann hypothesis, as several of its "predictions" turned out to be true.
- Lehmer's phenomenon (Lehmer 1956) where two zeros are sometimes very close is sometimes given as a reason to disbelieve in the Riemann hypothesis. However one would expect this to happen occasionally just by chance even if the Riemann hypothesis were true, and Odlyzko's calculations suggest that nearby pairs of zeros occur just as often as predicted by Montgomery's conjecture.
- Patterson (1988) suggests that the most compelling reason for the Riemann hypothesis for most mathematicians is the hope that primes are distributed as regularly as possible.

Other Results

The location of the Riemann zeta function's zeros is of great importance in the theory of numbers. The prime number theorem is equivalent to the fact that there are no zeros of the zeta function on the Re(s) = 1 line. A better result is that $\zeta(\sigma + it) \neq 0$ whenever $|t| \geq 3$ and

$$\sigma \geq 1 - \frac{1}{57.54(\log|t|)^{2/3}(\log\log|t|)^{1/3}}.$$

The strongest result of this kind one can hope for is the truth of the Riemann hypothesis, which would have many profound consequences in the theory of numbers.

It is known that there are infinitely many zeros on the critical line. Littlewood showed that if the sequence (γ_n) contains the imaginary parts of all zeros in the upper half-plane in ascending order, then

$$\lim_{n\to\infty}\left(\gamma_{n+1}-\gamma_n\right)=0.$$

The critical line theorem asserts that a positive percentage of the nontrivial zeros lies on the critical line.

In the critical strip, the zero with smallest non-negative imaginary part is 1/2 + i14.13472514... (A058303). Directly from the functional equation one sees that the non-trivial zeros are symmetric about the axis Re(s) = 1/2. Furthermore, the fact that $\zeta(s)=\overline{\zeta(\overline{s})}$ for all complex $s \neq 1$ implies that the zeros of the Riemann zeta function are symmetric about the real axis.

The statistics of the Riemann zeta zeros are a topic of interest to mathematicians because of their connection to big problems like the Riemann hypothesis, distribution of prime numbers, etc. Through connections with random matrix theory and quantum chaos, the appeal is even broader. The fractal structure of the Riemann zeta zero distribution has been studied using rescaled range analysis. The self-similarity of the zero distribution is quite remarkable, and is characterized by a large fractal dimension of 1.9. This rather large fractal dimension is found over zeros covering at least fifteen orders of magnitude, and also for the zeros of other L-functions.

Various Properties

For sums involving the zeta-function at integer and half-integer values.

Reciprocal

The reciprocal of the zeta function may be expressed as a Dirichlet series over the Möbius function $\mu(n)$:

$$\frac{1}{\zeta(s)}=\sum_{n=1}^{\infty}\frac{\mu(n)}{n^s}$$

for every complex number s with real part > 1. There are a number of similar relations involving various well-known multiplicative functions; these are given in the article on the Dirichlet series.

The Riemann hypothesis is equivalent to the claim that this expression is valid when the real part of s is greater than 1/2.

Universality

The critical strip of the Riemann zeta function has the remarkable property of universality. This zeta-function universality states that there exists some location on the critical strip that approximates any holomorphic function arbitrarily well. Since holomorphic functions are very general, this property is quite remarkable.

Generalizations

There are a number of related zeta functions that can be considered to be generalizations of the Riemann zeta function. These include the Hurwitz zeta function

$$\zeta(s,q) = \sum_{k=0}^{\infty} \frac{1}{(k+q)^s}$$

(the convergent series representation was given by Helmut Hasse in 1930, cf. Hurwitz zeta function), which coincides with the Riemann zeta function when $q = 1$ (note that the lower limit of summation in the Hurwitz zeta function is 0, not 1), the Dirichlet L-functions and the Dedekind zeta-function.

The polylogarithm is given by

$$\mathrm{Li}_s(z) = \sum_{k=1}^{\infty} \frac{z^k}{k^s}$$

which coincides with the Riemann zeta function when $z = 1$.

The Lerch transcendent is given by

$$\Phi(z,s,q) = \sum_{k=0}^{\infty} \frac{z^k}{(k+q)^s}$$

which coincides with the Riemann zeta function when $z = 1$ and $q = 1$ (note that the lower limit of summation in the Lerch transcendent is 0, not 1).

The Clausen function $\mathrm{Cl}_s(\theta)$ that can be chosen as the real or imaginary part of $\mathrm{Li}_s(e^{i\theta})$.

The multiple zeta functions are defined by

$$\zeta(s_1, s_2, \ldots, s_n) = \sum_{k_1 > k_2 > \cdots > k_n > 0} k_1^{-s_1} k_2^{-s_2} \cdots k_n^{-s_n}.$$

One can analytically continue these functions to the n-dimensional complex space. The special values of these functions are called multiple zeta values by number theorists and have been connected to many different branches in mathematics and physics.

Faà di Bruno's Formula

Faà di Bruno's formula is an identity in mathematics generalizing the chain rule to higher derivatives, named after Francesco Faà di Bruno (1855, 1857), though he was not the first to state or prove the formula.

In 1800, more than 50 years before Faà di Bruno, the French mathematician Louis François Antoine Arbogast stated the formula in a calculus textbook, considered the first published reference on the subject.

Perhaps the most well-known form of Faà di Bruno's formula says that

$$\frac{d^n}{dx^n} f(g(x)) = \sum \frac{n!}{m_1!1!^{m_1} m_2!2!^{m_2} \cdots m_n!n!^{m_n}} \cdot f^{(m_1+\cdots+m_n)}(g(x)) \cdot \prod_{j=1}^{n} \left(g^{(j)}(x)\right)^{m_j},$$

where the sum is over all n-tuples of nonnegative integers (m_1, ..., m_n) satisfying the constraint

$$1 \cdot m_1 + 2 \cdot m_2 + 3 \cdot m_3 + \cdots + n \cdot m_n = n.$$

Sometimes, to give it a memorable pattern, it is written in a way in which the coefficients that have the combinatorial interpretation discussed below are less explicit:

$$\frac{d^n}{dx^n} f(g(x)) = \sum \frac{n!}{m_1!m_2!\cdots m_n!} \cdot f^{(m_1+\cdots+m_n)}(g(x)) \cdot \prod_{j=1}^{n} \left(\frac{g^{(j)}(x)}{j!}\right)^{m_j}.$$

Combining the terms with the same value of $m_1 + m_2 + \ldots + m_n = k$ and noticing that m_j has to be zero for $j > n - k + 1$ leads to a somewhat simpler formula expressed in terms of Bell polynomials $B_{n,k}(x_1, \ldots, x_{n-k+1})$:

$$\frac{d^n}{dx^n} f(g(x)) = \sum_{k=1}^{n} f^{(k)}(g(x)) \cdot B_{n,k}\left(g'(x), g''(x), \dots, g^{(n-k+1)}(x)\right).$$

Combinatorial Form

The formula has a "combinatorial" form:

$$\frac{d^n}{dx^n} f(g(x)) = (f \circ g)^{(n)}(x) = \sum_{\pi \in \Pi} f^{(|\pi|)}(g(x)) \cdot \prod_{B \in \pi} g^{(|B|)}(x)$$

$$\frac{d^n}{dx^n} f(g(x)) = (f \circ g)^{(n)}(x) = \sum_{\pi \in \Pi} f^{(|\pi|)}(g(x)) \cdot \prod_{B \in \pi} g^{(|B|)}(x)$$

where

- π runs through the set Π of all partitions of the set $\{ 1, \dots, n \}$,
- "$B \in \pi$" means the variable B runs through the list of all of the "blocks" of the partition π, and
- $|A|$ denotes the cardinality of the set A (so that $|\pi|$ is the number of blocks in the partition π and $|B|$ is the size of the block B).

Explication via an Example

The combinatorial form may initially seem forbidding, so let us examine a concrete case, and see what the pattern is:

$$\begin{aligned}(f \circ g)''''(x) &= f''''(g(x))g'(x)^4 + 6f'''(g(x))g''(x)g'(x)^2 \\ &\quad + 3f''(g(x))g''(x)^2 + 4f''(g(x))g'''(x)g'(x) \\ &\quad + f'(g(x))g''''(x).\end{aligned}$$

The pattern is

$g'(x)^4$	$\leftrightarrow$	$1+1+1+1$	$\leftrightarrow$	$f''''(g(x))$	$\leftrightarrow$	1
$g''(x)g'(x)^2$	$\leftrightarrow$	$2+1+1$	$\leftrightarrow$	$f'''(g(x))$	$\leftrightarrow$	6
$g''(x)^2$	$\leftrightarrow$	$2+2$	$\leftrightarrow$	$f''(g(x))$	$\leftrightarrow$	3
$g'''(x)g'(x)$	$\leftrightarrow$	$3+1$	$\leftrightarrow$	$f''(g(x))$	$\leftrightarrow$	4
$g''''(x)$	$\leftrightarrow$	4	$\leftrightarrow$	$f'(g(x))$	$\leftrightarrow$	1.

The factor $g''(x)g'(x)^2$ corresponds to the partition 2 + 1 + 1 of the integer 4, in the obvious way. The factor $f'''(g(x))$ that goes with it corresponds to the fact that there are *three* summands in that partition. The coefficient 6 that goes with those factors corresponds to the fact that there are exactly six partitions of a set of four members that break it into one part of size 2 and two parts of size 1.

Similarly, the factor $g''(x)^2$ in the third line corresponds to the partition 2 + 2 of the integer 4, (4, because we are finding the fourth derivative), while $f''(g(x))$ corresponds to the fact that there are *two* summands (2 + 2) in that partition. The coefficient 3 corresponds to the fact that there are $\frac{1}{2}\binom{4}{2} = 3$ ways of partitioning 4 objects into groups of 2. The same concept applies to the others.

Combinatorics of the Faà di Bruno Coefficients

These partition-counting Faà di Bruno coefficients have a "closed-form" expression. The number of partitions of a set of size n corresponding to the integer partition

$$n = \underbrace{1+\cdots+1}_{m_1} + \underbrace{2+\cdots+2}_{m_2} + \underbrace{3+\cdots+3}_{m_3} + \cdots$$

of the integer n is equal to

$$\frac{n!}{m_1!m_2!m_3!\cdots 1!^{m_1} 2!^{m_2} 3!^{m_3}\cdots}.$$

These coefficients also arise in the Bell polynomials, which are relevant to the study of cumulants.

Variations

Multivariate Version: Let $y = g(x_1,\ldots, x_n)$. Then the following identity holds regardless of whether the n variables are all distinct, or all identical, or partitioned into several distinguishable classes of indistinguishable variables:

$$\frac{\partial^n}{\partial x_1 \cdots \partial x_n} f(y) = \sum_{\pi\in\Pi} f^{(|\pi|)}(y) \cdot \prod_{B\in\pi} \frac{\partial^{|B|} y}{\prod_{j\in B} \partial x_j}$$

where (as above)

- π runs through the set Π of all partitions of the set $\{ 1,\ldots, n \}$,

- "$B \in \pi$" means the variable B runs through the list of all of the "blocks" of the partition π, and
- $|A|$ denotes the cardinality of the set A (so that $|\pi|$ is the number of blocks in the partition π and $|B|$ is the size of the block B).

A further generalization, due to Tsoy-Wo Ma, considers the case where y is a vector-valued variable. The general form, for variational calculus (Gateaux differentials are the most general form of differential), was derived in 2012.

Example

The five terms in the following expression correspond in the obvious way to the five partitions of the set $\{1, 2, 3\}$, and in each case the order of the derivative of f is the number of parts in the partition:

$$\begin{aligned}\frac{\partial^3}{\partial x_1 \partial x_2 \partial x_3} f(y) &= f'(y)\frac{\partial^3 y}{\partial x_1 \partial x_2 \partial x_3} \\ &+ f''(y)\left(\frac{\partial y}{\partial x_1}\cdot\frac{\partial^2 y}{\partial x_2 \partial x_3} + \frac{\partial y}{\partial x_2}\cdot\frac{\partial^2 y}{\partial x_1 \partial x_3} + \frac{\partial y}{\partial x_3}\cdot\frac{\partial^2 y}{\partial x_1 \partial x_2}\right) \\ &+ f'''(y)\frac{\partial y}{\partial x_1}\cdot\frac{\partial y}{\partial x_2}\cdot\frac{\partial y}{\partial x_3}.\end{aligned}$$

If the three variables are indistinguishable from each other, then three of the five terms above are also indistinguishable from each other, and then we have the classic one-variable formula.

Formal Power Series Version

In the formal power series

$$f(x) = \sum_n \frac{a_n}{n!} x^n,$$

we have the nth derivative at 0:

$$f^{(n)}(0) = a_n.$$

This should not be construed as the value of a function, since these series are purely formal; there is no such thing as convergence or divergence in this context.

If

$$g(x) = \sum_{n=1}^{\infty} \frac{b_n}{n!} x^n$$

and

$$f(x) = \sum_{n=1}^{\infty} \frac{a_n}{n!} x^n$$

and

$$g(f(x)) = h(x) = \sum_{n=1}^{\infty} \frac{c_n}{n!} x^n,$$

then the coefficient c_n (which would be the nth derivative of h evaluated at 0 if we were dealing with convergent series rather than formal power series) is given by

$$c_n = \sum_{\pi=\{B_1,\ldots,B_k\}} a_{|B_1|} \cdots a_{|B_k|} b_k$$

where π runs through the set of all partitions of the set $\{1,\ldots, n\}$ and $B_1, \ldots, B_k$ are the blocks of the partition π, and $|B_j|$ is the number of members of the jth block, for $j = 1, \ldots, k$.

This version of the formula is particularly well suited to the purposes of combinatorics.

We can also write

$$g(f(x)) = \sum_{n=1}^{\infty} \frac{\sum_{k=1}^{n} b_k B_{n,k}(a_1,\ldots,a_{n-k+1})}{n!} x^n,$$

where $B_{n,k}(a_1,\ldots,a_{n"k+1})$ are Bell polynomials.

A Special Case

If $f(x) = e^x$ then all of the derivatives of f are the same, and are a factor common to every term. In case $g(x)$ is a cumulant-generating function, then $f(g(x))$ is a moment-generating function, and the polynomial in various derivatives of g is the polynomial that expresses the moments as functions of the cumulants.

Chapter 7

General Leibniz Rule

In calculus, the general Leibniz rule, named after Gottfried Leibniz, generalizes the product rule (which is also known as "Leibniz's rule") It states that if f and g are n-times differentiable functions, then the nth derivative of the product fg is given by

$$(f \cdot g)^{(n)} = \sum_{k=0}^{n} \binom{n}{k} f^{(k)} g^{(n-k)}$$

where $\binom{n}{k}$ is the binomial coefficient.

This can be proved by using the product rule and mathematical induction. With the multi-index notation the rule states more generally:

$$\partial^{\alpha}(fg) = \sum_{\{\beta:\beta\le\alpha\}} \binom{\alpha}{\beta} (\partial^{\alpha-\beta} f)(\partial^{\beta} g).$$

This formula can be used to derive a formula that computes the symbol of the composition of differential operators. In fact, let P and Q be differential operators (with coefficients that are differentiable sufficiently many times) and $R = P \circ Q$. Since R is also a differential operator, the symbol of R is given by:

$$R(x,\xi) = e^{-\langle x,\xi\rangle} R(e^{\langle x,\xi\rangle}).$$

A direct computation now gives:

$$R(x,\xi) = \sum_{\alpha} \frac{1}{\alpha!} \left(\frac{\partial}{\partial \xi}\right)^{\alpha} P(x,\xi) \left(\frac{\partial}{\partial x}\right)^{\alpha} Q(x,\xi).$$

This formula is usually known as the Leibniz formula. It is used to define the composition in the space of symbols, thereby inducing the ring structure.

Integral

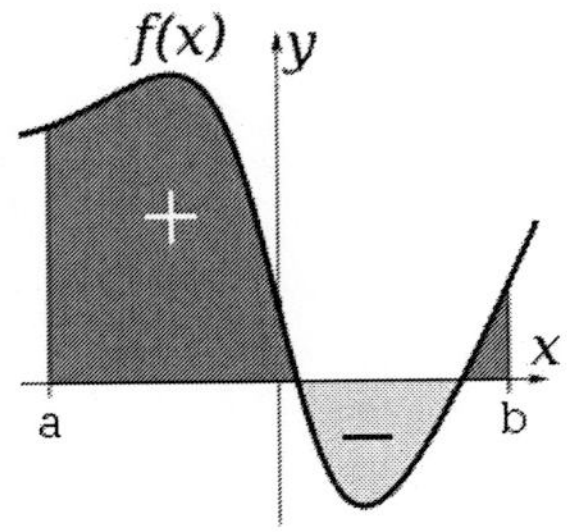

***Figure:** A definite integral of a function can be represented as the signed area of the region bounded by its graph.*

Integration is an important concept in mathematics and, together with its inverse, differentiation, is one of the two main operations in calculus. Given a function f of a real variable x and an interval $[a, b]$ of the real line, the definite integral

$$\int_a^b f(x)dx$$

is defined informally to be the area of the region in the xy-plane bounded by the graph of f, the x-axis, and the vertical lines $x = a$ and $x = b$, such that area above the x-axis adds to the total, and that below the x-axis subtracts from the total.

The term *integral* may also refer to the notion of the antiderivative, a function F whose derivative is the given function f. In this case, it is called an *indefinite integral* and is written:

$$F = \int f(x)dx.$$

The integrals discussed in this article are termed *definite integrals.*

The principles of integration were formulated independently by Isaac Newton and Gottfried Leibniz in the late 17th century. Through the fundamental theorem of calculus, which they independently developed, integration is connected with differentiation: if f is a continuous real-valued function defined on a closed interval $[a, b]$, then, once an antiderivative F of f is known, the definite integral of f over that interval is given by

$$\int_a^b f(x)dx = F(b) - F(a)$$

Integrals and derivatives became the basic tools of calculus, with numerous applications in science and engineering. The founders of the calculus thought of the integral as an infinite sum of rectangles of infinitesimal width. A rigorous mathematical definition of the integral was given by Bernhard Riemann. It is based on a limiting procedure which approximates the area of a curvilinear region by breaking the region into thin vertical slabs. Beginning in the nineteenth century, more sophisticated notions of integrals began to appear, where the type of the function as well as the domain over which the integration is performed has been generalised. A line integral is defined for functions of two or three variables, and the interval of integration $[a, b]$ is replaced by a certain curve connecting two points on the plane or in the space. In a surface integral, the curve is replaced by a piece of a surface in the three-dimensional space. Integrals of differential forms play a fundamental role in modern differential geometry. These generalizations of integrals first arose from the needs of physics, and they play an important role in the formulation of many physical laws, notably those of electrodynamics. There are many modern concepts of integration, among these, the most common is based on the abstract mathematical theory known as Lebesgue integration, developed by Henri Lebesgue.

Lebesgue Integration

In mathematics, the integral of a non-negative function can be regarded in the simplest case as the area between the graph of that function and the x-axis. Lebesgue integration is a mathematical construction that extends the integral to a larger class of functions; it also extends the domains on which these functions can be defined. It had long been understood that for non-negative functions with a smooth enough graph (such as continuous functions on closed bounded intervals) the *area under the curve* could be defined as the integral and computed using techniques of approximation of the region by polygons. However, as the need to consider more irregular functions arose (for example, as a result of the limiting processes of mathematical analysis and the mathematical theory of probability) it became clear that more careful approximation techniques would be needed to define a suitable integral. Also, we might wish to integrate on spaces more general than the real line; the Lebesgue integral provides the right abstractions needed to do this important job.

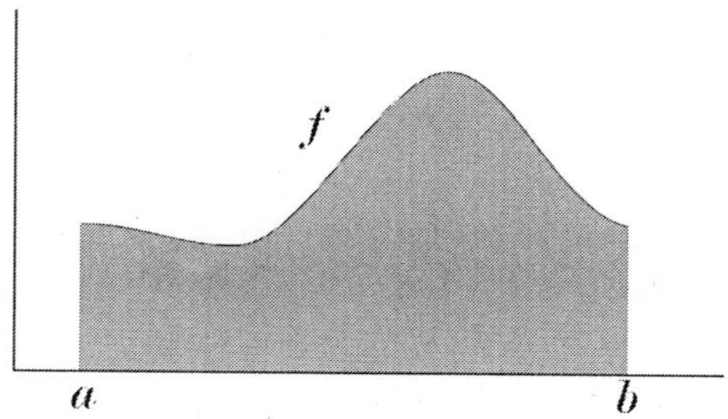

Figure: *The integral of a positive function can be interpreted as the area under a curve.*

The Lebesgue integral plays an important role in the branch of mathematics called real analysis and in many other fields in the mathematical sciences, and is named after Henri Lebesgue (1875–1941) who introduced the integral in (Lebesgue 1904). It is also a pivotal portion of the axiomatic theory of probability.

The term "Lebesgue integration" may refer either to the general theory of integration of a function with respect to a general measure, as introduced by Lebesgue, or to the specific case of integration of a function defined on a sub-domain of the real line with respect to Lebesgue measure.

The integral of a function f between limits a and b can be interpreted as the area under the graph of f. This is easy to understand for familiar functions such as polynomials, but what does it mean for more exotic functions? In general, what is the class of functions for which "area under the curve" makes sense? The answer to this question has great theoretical and practical importance.

As part of a general movement towards rigour in mathematics in the nineteenth century, attempts were made to put the integral calculus on a firm foundation. The Riemann integral, proposed by Bernhard Riemann (1826–1866), is a broadly successful attempt to provide such a foundation. Riemann's definition starts with the construction of a sequence of easily-calculated areas which converge to the integral of a given function. This definition is successful in the sense that it gives the expected answer for many already-solved problems, and gives useful results for many other problems.

However, Riemann integration does not interact well with taking limits of sequences of functions, making such limiting processes difficult to analyze. This is of prime importance, for instance, in the study of Fourier series, Fourier transforms and other topics.

The Lebesgue integral is better able to describe how and when it is possible to take limits under the integral sign. The Lebesgue definition considers a different class of easily-calculated areas than the Riemann definition, which is the main reason the Lebesgue integral is better behaved. The Lebesgue definition also makes it possible to calculate integrals for a broader class of functions. For example, the Dirichlet function, which is 0 where its argument is irrational and 1 otherwise, has a Lebesgue integral, but it does not have a Riemann integral.

Lebesgue's approach to integration was summarized in a letter to Paul Montel. He writes:

> *I have to pay a certain sum, which I have collected in my pocket. I take the bills and coins out of my pocket and give them to the creditor in the order I find them until I have reached the total sum. This is the Riemann integral. But I can proceed differently. After I have taken all the money out of my pocket I order the bills and coins according to identical values and then I pay the several heaps one after the other to the creditor. This is my integral.*
>
> *—Source: (Siegmund-Schultze 2008)*

The insight is that one should be able to rearrange the values of a function freely while preserving the value of the integral. This process of rearrangement can convert a very pathological function into one which is "nice" from the point of view of integration, and thus allows for such pathological functions to be integrated.

Intuitive Interpretation

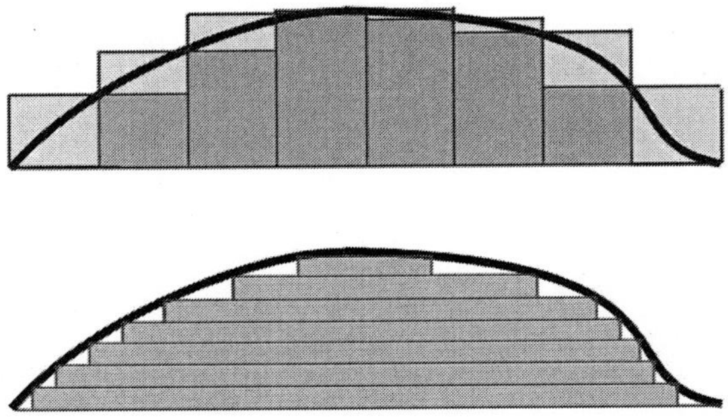

Figure: *Riemann-Darboux's integration (in blue) and Lebesgue integration (in red).*

To get some intuition about the different approaches to integration, let us imagine that it is desired to find a mountain's volume (above sea level).

The Riemann-Darboux Approach: Divide the base of the mountain into a grid of 1 metre squares. Measure the altitude of the mountain at the centre of each square. The volume on a single grid square is approximately 1×1×(altitude), so the total volume is the sum of the altitudes.

The Lebesgue Approach: Draw a contour map of the mountain, where adjacent contours are 1 metre of altitude apart. The volume of earth contained in a single contour is approximately that contour's area times 1, so the total volume is the sum of these areas times 1 metre.

Folland summarizes the difference between the Riemann and Lebesgue approaches thus: "to compute the Riemann integral of f, one partitions the domain $[a, b]$ into subintervals", while in the Lebesgue integral, "one is in effect partitioning the range of f".

Towards a Formal Definition

To define the Lebesgue integral formally requires the notion of a measure which, roughly, associates to each set A of real numbers a nonnegative number $\mu(A)$ representing the "size" of A. This notion of "size" should agree with the usual length of an interval or disjoint union of intervals. Suppose that $f : \mathbb{R} \to \mathbb{R}$ is a non-negative real-valued function. Using the "partitioning the range of f' philosophy, the integral of f should be the sum over t of the area of the thin horizontal strip between $y = t$ and $y = t+dt$. This area is just

$$\mu(\{x \mid f(x) > t\})dt.$$

Let

$$f^*(t) = \mu(\{x \mid f(x) > t\}).$$

The Lebesgue integral of f is then defined by

$$\int f = \int_0^{\infty} f^*(t)dt$$

where the integral on the right is an ordinary improper Riemann integral (note that f^* is a decreasing non-negative function, and therefore has a well-defined improper Riemann integral). For a suitable class of functions (the measurable functions) this defines the Lebesgue integral.

A general (not necessarily positive) function f is Lebesgue integrable if the area between the graph of f and the x-axis is finite:

$$\int |f| d\mu < +\infty.$$

In that case, the integral is, as in the Riemannian case, the difference between the area above the x-axis and the area below the x-axis:

$$\int f d\mu = \int f^+ d\mu - \int f^- d\mu$$

where

$$f^+(x) = \max(\{f(x), 0\}) = \begin{cases} f(x), & \text{if } f(x) > 0, \\ 0, & \text{otherwise} \end{cases}$$

$$f^-(x) = \max(\{-f(x), 0\}) = \begin{cases} -f(x), & \text{if } f(x) < 0, \\ 0, & \text{otherwise.} \end{cases}$$

Construction

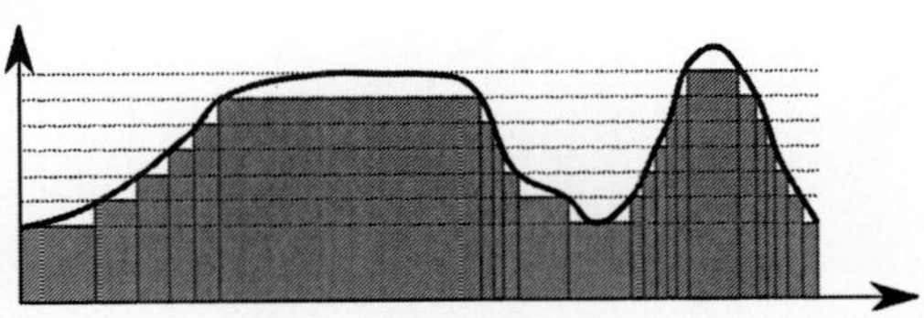

Figure: *Approximating a function by simple functions.*

The discussion that follows parallels the most common expository approach to the Lebesgue integral. In this approach, the theory of integration has two distinct parts:

1. A theory of measurable sets and measures on these sets.
2. A theory of measurable functions and integrals on these functions.

The function whose integral is to be found is then approximated by certain so-called simple functions, whose integrals can be written in terms of the measure. The integral of the original function is then the limit of the integral of the simple functions.

Measure Theory

Measure theory was initially created to provide a useful abstraction of the notion of length of subsets of the real line and, more generally, area and volume of subsets of Euclidean spaces. In particular, it provided a systematic answer to the question of which subsets of R have a length. As was shown by later developments in set theory, it is actually impossible to assign a length to all subsets of R in a way which preserves some natural additivity and translation invariance properties. This suggests that picking out a suitable class of *measurable* subsets is an essential prerequisite.

The Riemann integral uses the notion of length explicitly. Indeed, the element of calculation for the Riemann integral is the rectangle $[a, b] \times [c, d]$, whose area is calculated to be $(b - a)(d - c)$. The quantity $b - a$ is the length of the base of the rectangle and $d - c$ is the height of the rectangle. Riemann could only use planar rectangles to approximate the area under the curve because there was no adequate theory for measuring more general sets.

In the development of the theory in most modern textbooks (after 1950), the approach to measure and integration is *axiomatic*. This means that a measure is any function μ defined on a certain class X of subsets of a set E, which satisfies a certain list of properties. These properties can be shown to hold in many different cases.

Integration

We start with a measure space (E, X, μ) where E is a set, X is a σ-algebra of subsets of E and μ is a (non-negative) measure on E, defined on the sets of X.

For example, E can be Euclidean n-space R^n or some Lebesgue measurable subset of it, X will be the σ-algebra of all Lebesgue measurable subsets of E, and μ will be the Lebesgue measure. In the mathematical theory of probability, we confine our study to a probability measure μ, which satisfies $\mu(E) = 1$.

In Lebesgue's theory, integrals are defined for a class of functions called measurable functions. A function f is measurable if the pre-image of every interval of the form (t, ") is in X:

$$\{x|f(x) > t\} \in X \quad \text{for all } t \in \mathrm{R}.$$

It can be shown that this is equivalent to requiring that the pre-image of any Borel subset of R be in X. We will make this assumption henceforth. The set of measurable functions is closed under algebraic operations, but more importantly the class is closed under various kinds of pointwise sequential limits:

$$\sup_{k\in \mathrm{N}} f_k, \quad \liminf_{k\in \mathrm{N}} f_k, \quad \limsup_{k\in \mathrm{N}} f_k$$

are measurable if the original sequence $(f_k)_k$, where $k \in \mathrm{N}$, consists of measurable functions.

We build up an integral

$$\int_E f\,d\mu = \int_E f(x)\mu(dx)$$

for measurable real-valued functions f defined on E in stages:

Indicator Functions: To assign a value to the integral of the indicator function 1_S of a measurable set S consistent with the given measure μ, the only reasonable choice is to set:

$$\int 1_S\,d\mu = \mu(S).$$

Notice that the result may be equal to $+\infty$, unless μ is a *finite* measure.

Simple Functions: A finite linear combination of indicator functions

$$\sum_k a_k 1_{S_k}$$

where the coefficients a_k are real numbers and the sets S_k are measurable, is called a measurable simple function. We extend the integral by linearity to *non-negative* measurable simple functions. When the coefficients a_k are non-negative, we set

$$\int\left(\sum_k a_k 1_{S_k}\right)d\mu = \sum_k a_k \int 1_{S_k}\, d\mu = \sum_k a_k\, \mu(S_k).$$

The convention $0 \times \infty = 0$ must be used, and the result may be infinite. Even if a simple function can be written in many ways as a linear combination of indicator functions, the integral will always be the same; this can be shown using the additivity property of measures. Some care is needed when defining the integral of a *real-valued* simple function, to avoid the undefined expression $\infty - \infty$: one assumes that the representation

$$f = \sum_k a_k 1_{S_k}$$

is such that $\mu(S_k) < \infty$ whenever $a_k \neq 0$. Then the above formula for the integral of f makes sense, and the result does not depend upon the particular representation of f satisfying the assumptions.

If B is a measurable subset of E and s is a measurable simple function one defines

$$\int_B s\,d\mu = \int 1_B\, s\,d\mu = \sum_k a_k\, \mu(S_k \cap B).$$

Non-negative functions: Let f be a non-negative measurable function on E which we allow to attain the value $+\infty$, in other words, f takes non-negative values in the extended real number line. We define

$$\int_E f\,d\mu = \sup\left\{\int_E s\,d\mu : 0 \leq s \leq f,\, s \text{ simple}\right\}.$$

We need to show this integral coincides with the preceding one, defined on the set of simple functions. When E is a segment $[a, b]$, there is also the question of whether this corresponds in any way to a Riemann notion of integration. It is possible to prove that the answer to both questions is yes.

We have defined the integral of f for any non-negative extended real-valued measurable function on E. For some functions, this integral $+\infty_E f$ dμ will be infinite.

Signed Functions: To handle signed functions, we need a few more definitions. If f is a measurable function of the set E to the reals (including ±"), then we can write

$$f = f^{+} - f^{-},$$

where

$$f^{+}(x) = \begin{cases} f(x) & \text{if } f(x) > 0 \\ 0 & \text{otherwise} \end{cases}$$

$$f^{-}(x) = \begin{cases} -f(x) & \text{if } f(x) < 0 \\ 0 & \text{otherwise} \end{cases}$$

Note that both f^{+} and f^{-} are non-negative measurable functions. Also note that

$$|f| = f^{+} + f^{-}.$$

We say that the Lebesgue integral of the measurable function f *exists*, or *is defined* if at least one of $\int f^{+} d\mu$ and $\int f^{-} d\mu$ is finite:

$$\min\left(\int f^{+} d\mu, \int f^{-} d\mu\right) < \infty.$$

In this case we *define*

$$\int f d\mu = \int f^{+} d\mu - \int f^{-} d\mu.$$

If

$$\int |f| d\mu < \infty,$$

we say that f is *Lebesgue integrable.*

It turns out that this definition gives the desirable properties of the integral.

Complex valued functions can be similarly integrated, by considering the real part and the imaginary part separately.

Example

Consider the indicator function of the rational numbers, 1_Q. This function is nowhere continuous.

- 1_Q is not Riemann-integrable on [0,1]: No matter how the set [0,1] is partitioned into subintervals, each partition will contain at least one rational and at least one irrational number, because rationals and irrationals are both dense in the reals. Thus the upper Darboux sums will all be one, and the lower Darboux sums will all be zero.

- 1_Q is Lebesgue-integrable on [0,1] using the Lebesgue measure: Indeed it is the indicator function of the rationals so by definition

  $$\int_{[0,1]} 1_Q \, d\mu = \mu(Q \cap [0,1]) = 0,$$

  because Q is countable.

Domain of Integration

A technical issue in Lebesgue integration is that the domain of integration is defined as a *set* (a subset of a measure space), with no notion of orientation. In elementary calculus, one defines integration with respect to an orientation:

$$\int_b^a f := -\int_a^b f.$$

Generalizing this to higher dimensions yields integration of differential forms. By contrast, Lebesgue integration provides an alternative generalization, integrating over subsets with respect to a measure; this can be notated as

$$\int_A f \, d\mu = \int_{[a,b]} f \, d\mu$$

to indicate integration over a subset A.

Limitations of the Riemann Integral

Here we discuss the limitations of the Riemann integral and the greater scope offered by the Lebesgue integral. We presume a working understanding of the Riemann integral.

With the advent of Fourier series, many analytical problems involving integrals came up whose satisfactory solution required interchanging limit processes and integral signs. However, the conditions under which the integrals

$$\sum_k \int f_k(x) dx, \quad \int \left[\sum_k f_k(x) \right] dx$$

are equal proved quite elusive in the Riemann framework. There are some other technical difficulties with the Riemann integral. These are linked with the limit-taking difficulty discussed above.

Failure of monotone convergence. As shown above, the indicator function 1_Q on the rationals is not Riemann integrable. In particular, the Monotone convergence theorem fails. To see why, let $\{a_k\}$ be an enumeration of all the rational numbers in [0,1] (they are countable so this can be done.) Then let

$$g_k(x) = \begin{cases} 1 & \text{if } x = a_j, j \le k \\ 0 & \text{otherwise} \end{cases}$$

The function g_k is zero everywhere except on a finite set of points, hence its Riemann integral is zero. The sequence g_k is also clearly non-negative and monotonically increasing to 1_Q, which is not Riemann integrable.

Unsuitability for unbounded intervals. The Riemann integral can only integrate functions on a bounded interval. It can however be extended to unbounded intervals by taking limits, so long as this doesn't yield an answer such as $\infty-\infty$.

Integrating on structures other than Euclidean space. The Riemann integral is inextricably linked to the order structure of the line.

Alternative Formulations

It is possible to develop the integral with respect to the Lebesgue measure without relying on the full machinery of measure theory. One such approach is provided by the Daniell integral.

There is also an alternative approach to developing the theory of integration via methods of functional analysis. The Riemann integral exists for any continuous function f of compact support defined on R^n (or a fixed open subset). Integrals of more general functions can be built starting from these integrals.

Let C_c be the space of all real-valued compactly supported continuous functions of R. Define a norm on C_c by

$$\|f\| = \int |f(x)|\,dx.$$

Then C_c is a normed vector space (and in particular, it is a metric space.) All metric spaces have Hausdorff completions, so let L^1 be its completion. This space is isomorphic to the space of Lebesgue integrable functions modulo the subspace of functions with integral zero. Furthermore, the Riemann integral $+\infty$ is a uniformly continuous functional with respect to the norm on C_c, which is dense in L^1. Hence $+\infty$ has a unique extension to all of L^1. This integral is precisely the Lebesgue integral.

More generally, when the measure space on which the functions are defined is also a locally compact topological space (as is the case with the real numbers R), measures compatible with the topology in a suitable sense (Radon measures, of which the Lebesgue measure is an example) an integral with respect to them can be defined in the same manner, starting from the integrals of continuous functions with compact support. More precisely, the compactly supported functions form a vector space that carries a natural topology, and a (Radon) measure is defined as a continuous linear functional on this space. The value of a measure at a compactly supported function is then also by definition the integral of the function. One then proceeds to expand the measure (the integral) to more general functions by continuity, and defines the measure of a set as the integral of its indicator function. This is the approach taken by Bourbaki (2004) and a certain number of other authors.

Limitations of Lebesgue Integral

The main purpose of Lebesgue integral is to provide an integral notation where limits of integrals hold under mild assumptions. There is no guarantee that every function is Lebesgue integrable. It may happen that improper (Riemann) integrals exist for functions that are not Lebesgue integrable. One example would be

$$\frac{\sin(x)}{x}.$$

This function is not Lebesgue integrable as

$$\int \left|\frac{\sin(x)}{x}\right| d\mu = \infty.$$

On the other hand, it exists as an improper Riemann integral and the integral can be computed to be finite.

Multiple Integral

The multiple integral is a type of *definite* integral extended to functions of more than one real variable, for example, $f(x, y)$ or $f(x, y, z)$. Integrals of a function of two variables over a region in R^2 are called double integrals.

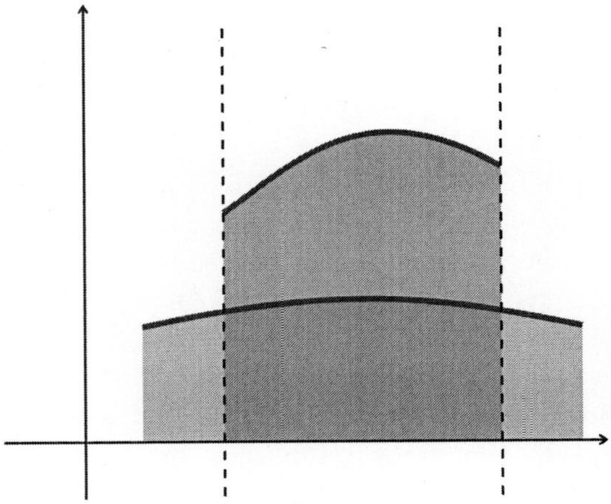

Figure: *Integral as area between two curves.*

Figure: *Double integral as volume under a surface* $z = x^2 - y^2$. *The rectangular region at the bottom of the body is the domain of integration, while the surface is the graph of the two-variable function to be integrated.*

Just as the definite integral of a positive function of one variable represents the area of the region between the graph of the function and the x-axis, the double integral of a positive function of two variables represents the volume of the region between the surface defined by the function (on the three dimensional Cartesian plane where $z = f(x, y)$) and the plane which contains its domain. (the same volume can be obtained via the triple integral—the integral of a function in three variables—of the constant function $f(x, y, z) = 1$ over the above-mentioned region between the surface and the plane.) If there are more variables, a multiple integral will yield hypervolumes of multi-dimensional functions.

Multiple integration of a function in n variables: $f(x_1, x_2, ..., x_n)$ over a domain D is most commonly represented by nested integral signs in the reverse order of execution (the leftmost integral sign is computed last), followed by the function and integrand arguments in proper order (the integral with respect to the rightmost argument is computed last). The domain of integration is either represented symbolically for every argument over each integral sign, or is abbreviated by a variable at the rightmost integral sign:

$$\int \cdots \int_{\mathrm{D}} f(x_1, x_2, \ldots, x_n)\, dx_1 \cdots dx_n$$

Since the concept of an antiderivative is only defined for functions of a single real variable, the usual definition of the indefinite integral does not immediately extend to the multiple integral.

Mathematical Definition

For n > 1, consider a so-called "half-open" n-dimensional hyperrectangular domain T, defined as:

$$T = [a_1, b_1) \times [a_2, b_2) \times \cdots \times [a_n, b_n) \subseteq \mathrm{R}^n.$$

Partition each interval $[a_j, b_j)$ into a finite family I_j of non-overlapping subintervals $i_{j\alpha}$, with each subinterval closed at the left end, and open at the right end.

Then the finite family of subrectangles C given by

$$C = I_1 \times I_2 \times \cdots \times I_n$$

is a partition of T; that is, the subrectangles C_k are non-overlapping and their union is T.

Let $f: T \to \mathrm{R}$ be a function defined on T. Consider a partition C of T as defined above, such that C is a family of m subrectangles C_m and

$$T = C_1 \cup C_2 \cup \cdots \cup C_m$$

We can approximate the total nth-dimensional volume bounded below by T and above by f with the following Riemann sum:

$$\sum_{k=1}^{m} f(P_k)\mathrm{m}(C_k)$$

where P_k is a point in C_k and $\mathrm{m}(C_k)$ is the product of the lengths of the intervals whose Cartesian product is C_k, otherwise known as the measure of C_k.

The diameter of a subrectangle C_k is the largest of the lengths of the intervals whose Cartesian product is C_k. The diameter of a given partition of T is defined as the largest of the diameters of the subrectangles in the partition. Intuitively, as the diameter of the partition C is restricted smaller and smaller, the number of subrectangles m gets larger, and the measure $\mathrm{m}(C_k)$ of each subrectangle grows smaller. The function f is said to be Riemann integrable if the limit

$$S = \lim_{\delta \to 0} \sum_{k=1}^{m} f(P_k)\mathrm{m}(C_k)$$

exists, where the limit is taken over all possible partitions of T of diameter at most δ.

If f is Riemann integrable, S is called the Riemann integral of f over T and is denoted

$$\int \cdots \int_T f(x_1, x_2, \ldots, x_n)\, dx_1 \cdots dx_n$$

Frequently this notation is abbreviated as

$$\int_T f(\mathrm{x}) d\mathrm{x}.$$

where x represents the n-tuple $(x_1, \ldots x_n)$ and dx is the n-dimensional volume differential.

The Riemann integral of a function defined over an arbitrary bounded n-dimensional set can be defined by extending that function to a function defined over a half-open rectangle whose values are zero outside the domain of the original function. Then the integral of the original function over the original domain is defined to be the integral of the extended function over its rectangular domain, if it exists.

In what follows the Riemann integral in n dimensions will be called multiple integral.

Properties

Multiple integrals have many properties common to those of integrals of functions of one variable (linearity, commutativity, monotonicity, and so on.). One important property of multiple integrals is that the value of an integral is independent of the order of integrands under certain conditions. This property is popularly known as Fubini's theorem.

Particular Cases

In the case of $T \subseteq R^2$, the integral

$$\ell = \iint_T f(x,y)\,dx\,dy$$

is the double integral of f on T, and if $T \subseteq R^3$ the integral

$$\ell = \iiint_T f(x,y,z)\,dx\,dy\,dz$$

is the triple integral of f on T.

Notice that, by convention, the double integral has two integral signs, and the triple integral has three; this is a notational convention which is convenient when computing a multiple integral as an iterated integral, as shown later in this article.

Methods of Integration

The resolution of problems with multiple integrals consists, in most of cases, of finding a way to reduce the multiple integral to an iterated integral, a series of integrals of one variable, each being directly solvable. Sometimes, it is possible to obtain the result of the integration by direct examination without any calculations.

Integrating Constant Functions

When the integrand is a constant function c, the integral is equal to the product of c and the measure of the domain of integration. If c = 1 and the domain is a subregion of R^2, the integral gives the area of the region, while if the domain is a subregion of R^3, the integral gives the volume of the region.

Example. Let $f(x, y)$ = 2 and

$$D = \{(x, y) \in \mathrm{R}^2 \leq : 2 \leq x \leq 4 \leq ; 3 \leq y \leq 6\}$$

in which case

$$\int_3^6 \int_2^4 2\, dx dy = 2\int_3^6 \int_2^4 1\, dx dy = 2\cdot\text{area}(D) = (2\cdot 3)\cdot 2 = 12,$$

since by definition we have:

$$\int_3^6 \int_2^4 1\, dx dy = \text{area}(D).$$

Use of Symmetry

When the domain of integration is symmetric about the origin with respect to at least one of the variables of integration and the integrand is odd with respect to this variable, the integral is equal to zero, as the integrals over the two halves of the domain have the same absolute value but opposite signs. When the integrand is even with respect to this variable, the integral is equal to twice the integral over one half of the domain, as the integrals over the two halves of the domain are equal.

Example 1: Consider the function $f(x, y) = 2\sin(x) - 3y^3 + 5$ integrated over the domain

$$T = \left\{(x, y) \in \mathrm{R}^2 \leq : x^2 + y^2 \leq 1\right\},$$

a disc with radius 1 centred at the origin with the boundary included. Using the linearity property, the integral can be decomposed into three pieces:

$$\iint_T (2\sin x - 3y^3 + 5) dx dy = \iint_T 2\sin x\, dx dy - \iint_T 3y^3 dx dy + \iint_T 5 dx dy$$

$2\sin(x)$ and $3y^3$ are both odd functions and moreover it is evident that the T disc has a symmetry for the x and even the y axis; therefore the only contribution to the final result of the integrals is that of the constant function 5 because the other two pieces are null.

Example 2: Consider the function $f(x, y, z) = x \exp(y^2 + z^2)$ and as integration region the sphere with radius 2 centred at the origin,

$$T = \left\{(x, y, z) \in \mathrm{R}^3 \ : x^2 + y^2 + z^2 \leq 4\right\}.$$

The "ball" is symmetric about all three axes, but it is sufficient to integrate with respect to x-axis to show that the integral is 0, because the function is an odd function of that variable.

Normal Domains on R²

This method is applicable to any domain D for which:

- the projection of D onto either the x-axis or the y-axis is bounded by the two values, a and b
- any line perpendicular to this axis that passes between these two values intersects the domain in an interval whose endpoints are given by the graphs of two functions, α and β.

x-axis

If the domain D is normal with respect to the x-axis, and $f : D \to \mathrm{R}$ is a continuous function; then $\alpha(x)$ and $\beta(x)$ (defined on the interval $[a, b]$) are the two functions that determine D. Then:

$$\iint_D f(x, y)\, dx dy = \int_a^b dx \int_{\alpha(x)}^{\beta(x)} f(x, y) dy.$$

y-axis

If D is normal with respect to the y-axis and $f : D \to \mathrm{R}$ is a continuous function; then $\alpha(y)$ and $\beta(y)$ (defined on the interval $[a, b]$) are the two functions that determine D. Then:

$$\iint_D f(x, y)\, dx dy = \int_a^b dy \int_{\alpha(y)}^{\beta(y)} f(x, y) dx.$$

Examples

Double Integral

Let us assume that we wish to integrate a multivariable function f over a region A.

$A = \left\{(x, y) \in \mathrm{R}^2 \ : 11 \leq x \leq 14 \ ; 7 \leq y \leq 10\right\}$ and $f(x, y) = x^2 + 4y$

From this we formulate the double integral

$$\int_7^{10} \int_{11}^{14} (x^2 + 4y)\, dx dy$$

The inner integral is performed first, integrating with respect to x and taking y as a constant, as it is not the variable of integration. The result of this integral, which is a function depending only on y, is then integrated with respect to y.

$$\begin{aligned}\int_{11}^{14}(x^2+4y)\,dx &= \left(\frac{1}{3}x^3+4yx\right)\Big|_{x=11}^{x=14}\\ &= \frac{1}{3}(14)^3+4y(14)-\frac{1}{3}(11)^3-4y(11)\\ &= 471+12y\end{aligned}$$

We then integrate the result with respect to y.

$$\begin{aligned}\int_{7}^{10}(471+12y)\,dy &= (471y+6y^2)\Big|_{y=7}^{y=10}\\ &= 471(10)+6(10)^2-471(7)-6(7)^2\\ &= 1719\end{aligned}$$

Observe that the order of integration is sometimes interchangeable:

$$\begin{aligned}\int_{11}^{14}\int_{7}^{10}(x^2+4y)\,dydx &= \int_{11}^{14}\left(x^2y+2y^2\right)\Big|_{y=7}^{y=10}dx\\ &= \int_{11}^{14}(3x^2+102)\,dx\\ &= \left(x^3+102x\right)\Big|_{x=11}^{x=14}\\ &= 1719\end{aligned}$$

The instances where the order is interchangeable is determined by Fubini's Theorem.

Volumes

The volume of the parallelepiped of sides 4 × 6 × 5 may be obtained in two ways:

- By calculating the double integral of the function $f(x, y) = 5$ over the region D in the xy-plane which is the base of the parallelepiped.

  $$\iint_D 5\,dxdy$$

- By calculating the triple integral of the constant function 1 over the parallelepiped itself

$$\iiint_{\text{parallelepiped}} 1 dx dy dz$$

Computing a Volume

Using the methods previously described, it is possible to calculate the volumes of some common solids.

- Cylinder: The volume of a cylinder with height h and circular base of radius R can be calculated by integrating the constant function h over the circular base, using polar coordinates.

$$\text{Volume} = \int_0^{2\pi} d\phi \int_0^R h\rho \, d\rho = h2\pi \left[\frac{\rho^2}{2} \right]_0^R = \pi R^2 h$$

This is in agreement with the formula

$$\text{Volume} = \text{base area} \times height .$$

- Sphere: The volume of a sphere with radius R can be calculated by integrating the constant function 1 over the sphere, using spherical coordinates.

$$\begin{aligned}
\text{Volume} &= \iiint_D f(x,y,z) dx dy dz \\
&= \iiint_D 1 dV \\
&= \iiint_S \rho^2 \sin\phi d\rho d\theta d\phi \\
&= \int_0^{2\pi} d\theta \int_0^{\pi} \sin\phi d\phi \int_0^R \rho^2 \, d\rho \\
&= 2\pi \int_0^{\pi} \sin\phi d\phi \int_0^R \rho^2 \, d\rho \\
&= 2\pi \int_0^{\pi} \sin\phi \frac{R^3}{3} d\phi \\
&= \frac{2}{3} \pi R^3 [-\cos\phi]_0^{\pi} = \frac{4}{3} \pi R^3 .
\end{aligned}$$

- Tetrahedron (triangular pyramid or 3-simplex): The volume of a tetrahedron with its apex at the origin and edges of length l along the x, y and z axes can be calculated by integrating the constant function 1 over the tetrahedron.

$$
\begin{aligned}
\text{Volume} &= \int_0^\ell dx \int_0^{\ell-x} dy \int_0^{\ell-x-y} dz \\
&= \int_0^\ell dx \int_0^{\ell-x} (\ell - x - y) dy \\
&= \int_0^\ell \left[\ell^2 - 2\ell x + x^2 - \frac{(\ell - x)^2}{2} \right] dx \\
&= \ell^3 - \ell \ell^2 + \frac{\ell^3}{3} - \left[\frac{\ell^2 x}{2} - \frac{\ell x^2}{2} + \frac{x^3}{6} \right]_0^\ell \\
&= \frac{\ell^3}{3} - \frac{\ell^3}{6} = \frac{\ell^3}{6}
\end{aligned}
$$

This is in agreement with the formula

$$\text{Volume} = \frac{1}{3} \times \text{base area} \times \text{height} = \frac{1}{3} \times \frac{\ell^2}{2} \times \ell = \frac{\ell^3}{6}.$$

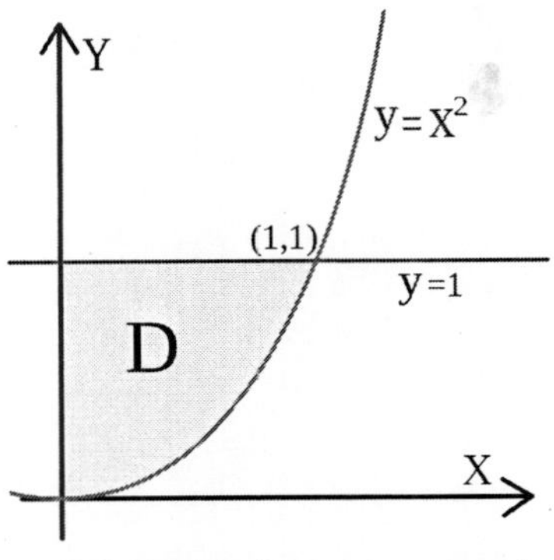

***Figure:** Example of an improper domain.*

Multiple Improper Integral

In case of unbounded domains or functions not bounded near the boundary of the domain, we have to introduce the double improper integral or the triple improper integral.

Multiple Integrals and Iterated Integrals

Fubini's theorem states that if

$$\int_{A \times B} |f(x, y)|\, d(x, y) < \infty,$$

that is, if the integral is absolutely convergent, then the multiple integral will give the same result as the iterated integral,

$$\int_{A\times B} f(x,y)d(x,y) = \int_A\left(\int_B f(x,y)dy\right)dx = \int_B\left(\int_A f(x,y)dx\right)dy.$$

In particular this will occur if $|f(x, y)|$ is a bounded function and A and B are bounded sets.

If the integral is not absolutely convergent, care is needed not to confuse the concepts of *multiple integral* and *iterated integral*, especially since the same notation is often used for either concept. The notation

$$\int_0^1\int_0^1 f(x,y)dydx$$

means, in some cases, an iterated integral rather than a true double integral. In an iterated integral, the outer integral

$$\int_0^1 \cdots dx$$

is the integral with respect to x of the following function of x:

$$g(x) = \int_0^1 f(x,y)dy.$$

A double integral, on the other hand, is defined with respect to area in the xy-plane. If the double integral exists, then it is equal to each of the two iterated integrals (either "$dy\ dx$" or "$dx\ dy$") and one often computes it by computing either of the iterated integrals. But sometimes the two iterated integrals exist when the double integral does not, and in some such cases the two iterated integrals are different numbers, i.e., one has

$$\int_0^1\int_0^1 f(x,y)dydx \neq \int_0^1\int_0^1 f(x,y)dxdy.$$

This is an instance of rearrangement of a conditionally convergent integral.

The notation

$$\int_{[0,1]\times[0,1]} f(x,y)dxdy$$

may be used if one wishes to be emphatic about intending a double integral rather than an iterated integral.

Some Practical Applications

Quite generally, just as in one variable, one can use the multiple integral to find the average of a function over a given set. Given a set $D \subseteq \mathrm{R}^n$ and an integrable function f over D, the average value of f over its domain is given by

$$\bar{f} = \frac{1}{m(D)} \int_D f(x)dx,$$

where $m(D)$ is the measure of D.

Additionally, multiple integrals are used in many applications in physics. The examples below also show some variations in the notation.

In mechanics, the moment of inertia is calculated as the volume integral (triple integral) of the density weighed with the square of the distance from the axis:

$$I_z = \iiint_V \rho r^2 \, dV.$$

The gravitational potential associated with a mass distribution given by a mass measure dm on three-dimensional Euclidean space R^3 is

$$V(\mathrm{x}) = -\int_{\mathrm{R}^3} \frac{G}{|\mathrm{x}-\mathrm{y}|} dm(\mathrm{y}).$$

If there is a continuous function $\rho(\mathrm{x})$ representing the density of the distribution at x, so that $dm(\mathrm{x}) = \rho(\mathrm{x})d$
$^3\mathrm{x}$, where d
$^3\mathrm{x}$ is the Euclidean volume element, then the gravitational potential is

$$V(\mathrm{x}) = -\int_{\mathrm{R}^3} \frac{G}{|\mathrm{x}-\mathrm{y}|} \rho(\mathrm{y}) d^3\mathrm{y}.$$

In electromagnetism, Maxwell's equations can be written using multiple integrals to calculate the total magnetic and electric fields. In the following example, the electric field produced by a distribution of charges given by the volume charge density $\rho(\vec{r})$ is obtained by a *triple integral* of a vector function:

$$\vec{E} = \frac{1}{4\pi\epsilon_0} \iiint \frac{\vec{r}-\vec{r}'}{\left\|\vec{r}-\vec{r}'\right\|^3} \rho(\vec{r}')\mathrm{d}^3 r'.$$

This can also be written as an integral with respect to a signed measure representing the charge distribution.

Iterated Integral

In calculus an iterated integral is the result of applying integrals to a function of more than one variable (for example $f(x,y)$ or $f(x,y,z)$) in a way that each of the integrals considers some of the variables as given constants. For example, the function $f(x,y)$,y if is considered a given parameter can be integrated with respect to x, $\int f(x,y)dx$. The result is a function of y and therefore its integral can be considered. If this is done, the result is the iterated integral

$$\int\left(\int f(x,y)dx\right)dy.$$

It is key for the notion of iterated integral that this is different, in principle, to the multiple integral

$$\iint f(x,y)dxdy.$$

Although in general these two can be different there is a theorem that, under very mild conditions, gives the equality of the two. This is Fubini's theorem.

The alternative notation for iterated integrals

$$\int dy\int f(x,y)dx$$

is also used. Iterated integrals are computed following the operational order indicated by the parentheses (in the notation that uses them). Starting from the most inner integral outside.

Examples

A simple computation:

For the iterated integral

$$\int\left(\int (x+y)dx\right)dy$$

the integral

$$\int (x+y)dx = \frac{x^2}{2} + yx$$

is computed first and then the result is used to compute the integral with respect to y.

$$\int (\frac{x^2}{2} + yx)dy = \frac{yx^2}{2} + \frac{xy^2}{2}$$

It should be noted, however, that this example omits the constants of integration. After the first integration with respect to x, we would rigorously need to introduce a "constant" function of y.

That is, If we were to differentiate this function with respect to x, any terms containing only y would vanish, leaving the original integrand. Similarly for the second integral, we would introduce a "constant" function of x, because we have integrated with respect to y.

In this way, indefinite integration does not make very much sense for functions of several variables. While the antiderivatives of single variable functions differ at most by a constant, the antiderivatives of multivariable functions differ by at most unknown single-variable terms, which could have a drastic effect on the behaviour of the function.

The Order is Important

The order in which the integrals are computed is important in iterated integrals. Examples in which the different orders lead to different results are usually for complicated functions as the one that follows.

Let a sequence $0 < a_1 < a_2 < \cdots$, such that $a_n \to 1$. Let g_n be continuous functions not vanishing in the interval (a_n, a_{n+1}) and zero elsewhere, such that $\int_0^1 g_n = 1$ for every n. Define

$$f(x,y) = \sum_{n=0}^{\infty} (g_n(x) - g_{n+1}(x)) g_n(y).$$

In the previous sum, at each specific (x, y), at most one term is different from zero. For this function it happens that

$$\int_0^1 \left(\int_0^1 f(x,y)dy \right) dx = 1 \neq 0 = \int_0^1 \left(\int_0^1 f(x,y)dx \right) dy$$

Improper Integral

In calculus, an improper integral is the limit of a definite integral as an endpoint of the interval(s) of integration approaches either a specified real number or “ or • or, in some cases, as both endpoints approach limits. Such an integral is often written symbolically just like a standard definite integral, perhaps with *infinity* as a limit of integration. But that conceals the limiting process.

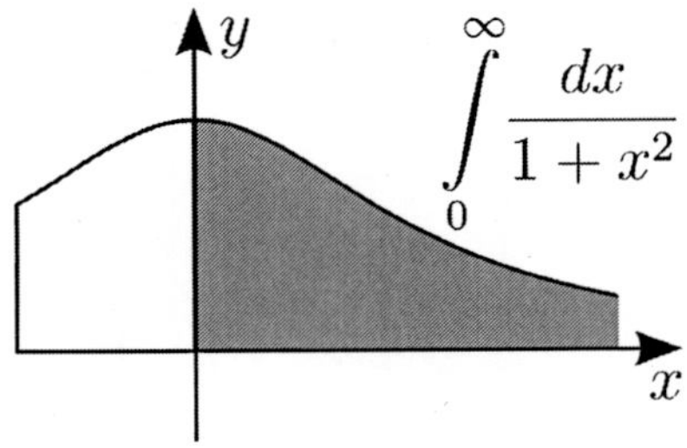

***Figure:** An improper integral of the first kind. The integral may need to be defined on an unbounded domain.*

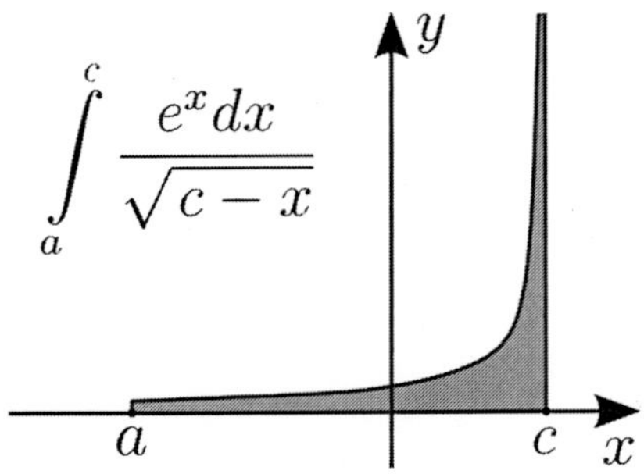

***Figure:** An improper Riemann integral of the second kind. The integral may fail to exist because of a vertical asymptote in the function.*

By using the more advanced Lebesgue integral, rather than the Riemann integral, one can in some cases get an answer without taking a limit of standard definite integrals.

Specifically, an improper integral is a limit of the form

$$\lim_{b\to\infty}\int_a^b f(x)\mathrm{d}x, \qquad \lim_{a\to-\infty}\int_a^b f(x)\mathrm{d}x,$$

or of the form

$$\lim_{c\to b^-}\int_a^c f(x)\mathrm{d}y, \quad \lim_{c\to a^+}\int_c^b f(x)\mathrm{d}x,$$

in which one takes a limit in one or the other (or sometimes both) endpoints (Apostol 1967, §10.23). Integrals are also improper if the integrand is undefined at an interior point of the domain of integration, or at multiple such points.

It is often necessary to use improper integrals in order to compute a value for integrals which may not exist in the conventional sense (as a Riemann integral, for instance) because of a singularity in the function, or an infinite endpoint of the domain of integration.

Examples

The original definition of the Riemann integral does not apply to a function such as $1/x^2$ on the interval [1, ∞), because in this case the domain of integration is unbounded. However, the Riemann integral can often be extended by continuity, by defining the improper integral instead as a limit

$$\int_1^\infty \frac{1}{x^2}\mathrm{d}x = \lim_{b\to\infty}\int_1^b \frac{1}{x^2}\mathrm{d}x = \lim_{b\to\infty}\left(-\frac{1}{b}+\frac{1}{1}\right) = 1.$$

The narrow definition of the Riemann integral also does not cover the function $1/\sqrt{x}$ on the interval [0, 1]. The problem here is that the integrand is unbounded in the domain of integration (the definition requires that both the domain of integration and the integrand be bounded). However, the improper integral does exist if understood as the limit

$$\int_0^1 \frac{1}{\sqrt{x}}\mathrm{d}x = \lim_{a\to 0^+}\int_a^1 \frac{1}{\sqrt{x}}\mathrm{d}x = \lim_{a\to 0^+}(2\sqrt{1}-2\sqrt{a}) = 2.$$

Convergence of the Integral

An improper integral converges if the limit defining it exists. Thus for example one says that the improper integral

$$\lim_{t\to\infty}\int_a^t f(x)\mathrm{d}x$$

exists and is equal to L if the integrals under the limit exist for all sufficiently large t, and the value of the limit is equal to L.

It is also possible for an improper integral to diverge to infinity. In that case, one may assign the value of " (or $-\infty$) to the integral. For instance

$$\lim_{b\to\infty}\int_1^b \frac{1}{x}\mathrm{d}x = \infty.$$

However, other improper integrals may simply diverge in no particular direction, such as

$$\lim_{b\to\infty}\int_1^b x\sin x\,dx,$$

which does not exist, even as an extended real number. This is called divergence by oscillation.

A limitation of the technique of improper integration is that the limit must be taken with respect to one endpoint at a time. Thus, for instance, an improper integral of the form

$$\int_{-\infty}^{\infty} f(x)dx$$

can be defined by taking two separate limits; to wit

$$\int_{-\infty}^{\infty} f(x)dx = \lim_{a\to-\infty}\lim_{b\to\infty}\int_a^b f(x)dx$$

provided the double limit is finite. It can also be defined as a pair of distinct improper integrals of the first kind:

$$\lim_{a\to-\infty}\int_a^c f(x)dx + \lim_{b\to\infty}\int_c^b f(x)dx$$

where c is any convenient point at which to start the integration. This definition also applies when one of these integrals is infinite, or both if they have the same sign.

An example of an improper integrals where both endpoints are infinite is the Gaussian integral $\int_{-\infty}^{\infty} e^{-x^2}dx = \sqrt{\pi}$. An example which evaluates to infinity is $\int_{-\infty}^{\infty} e^x dx$. But one cannot even define other integrals of this kind unambiguously, such as $\int_{-\infty}^{\infty} x dx$, since the double limit is infinite and the two-integral method

$$\lim_{a\to-\infty}\int_a^c x dx + \lim_{b\to\infty}\int_c^b x dx$$

yields $\infty - \infty$. In this case, one can however define an improper integral in the sense of Cauchy principal value:

$$\text{p.v.}\int_{-\infty}^{\infty} x dx = \lim_{b\to\infty}\int_{-b}^{b} x dx = 0.$$

The questions one must address in determining an improper integral are:

- Does the limit exist?
- Can the limit be computed?

The first question is an issue of mathematical analysis. The second one can be addressed by calculus techniques, but also in some cases by contour integration, Fourier transforms and other more advanced methods.

Types of Integrals

There is more than one theory of integration. From the point of view of calculus, the Riemann integral theory is usually assumed as the default theory. In using improper integrals, it can matter which integration theory is in play.

- For the Riemann integral (or the Darboux integral, which is equivalent to it), improper integration is necessary *both* for unbounded intervals (since one cannot divide the interval into finitely many subintervals of finite length) *and* for unbounded functions with finite integral (since, supposing it is unbounded above, then the upper integral will be infinite, but the lower integral will be finite).
- The Lebesgue integral deals differently with unbounded domains and unbounded functions, so that often an integral which only exists as an improper Riemann integral will exist as a (proper) Lebesgue integral, such as $\int_1^\infty \frac{1}{x^2}dx$. On the other hand, there are also integrals that have an improper Riemann integral but do not have a (proper) Lebesgue integral, such as $\int_0^\infty \frac{\sin x}{x}dx$. The Lebesgue theory does not see this as a deficiency: from the point of view of measure theory, $\int_0^\infty \frac{\sin x}{x}dx = \infty - \infty$ and cannot be defined satisfactorily. In some situations, however, it may be convenient to employ improper Lebesgue integrals as is the case, for instance, when defining the Cauchy principal value. The Lebesgue integral is more or less essential in the theoretical treatment of the Fourier transform, with pervasive use of integrals over the whole real line.
- For the Henstock–Kurzweil integral, improper integration *is not necessary*, and this is seen as a strength

of the theory: it encompasses all Lebesgue integrable and improper Riemann integrable functions.

Improper Riemann Integrals and Lebesgue Integrals

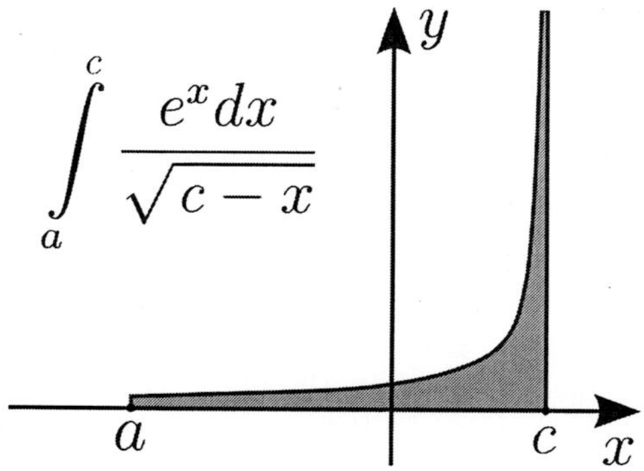

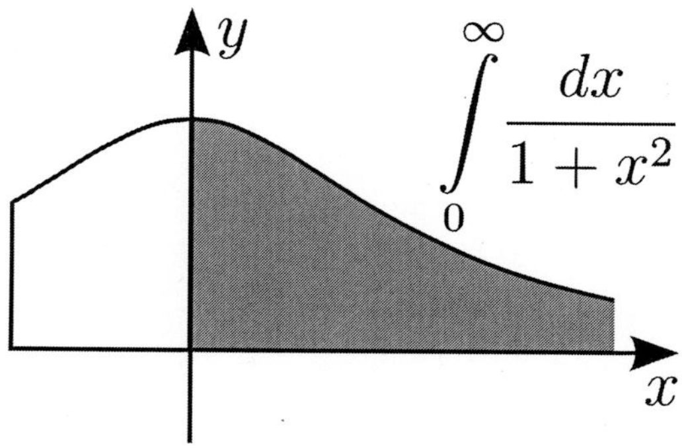

In some cases, the integral

$$\int_a^c f(x)\mathrm{d}x$$

can be defined as an integral (a Lebesgue integral, for instance) without reference to the limit

$$\lim_{b\to c^-}\int_a^b f(x)\mathrm{d}x$$

but cannot otherwise be conveniently computed. This often happens when the function f being integrated from a to c has a vertical asymptote at c, or if $c = \infty$.

In such cases, the improper Riemann integral allows one to calculate the Lebesgue integral of the function. Specifically, the following theorem holds (Apostol 1974, Theorem 10.33):

- If a function f is Riemann integrable on $[a,b]$ for every $b \geq a$, and the partial integrals

  $$\int_a^b |f(x)|\mathrm{d}x$$

  are bounded as $b \to \infty$, then the improper Riemann integrals

  $$\int_a^\infty f(x)\mathrm{d}x, \quad \text{and} \int_a^\infty |f(x)|\mathrm{d}x$$

  both exist. Furthermore, f is Lebesgue integrable on $[a, \infty)$, and its Lebesgue integral is equal to its improper Riemann integral.

For example, the integral

$$\int_0^\infty \frac{\mathrm{d}x}{1+x^2}$$

can be interpreted alternatively as the improper integral

$$\lim_{b\to\infty}\int_0^b \frac{\mathrm{d}x}{1+x^2} = \lim_{b\to\infty} \arctan b = \frac{\pi}{2},$$

or it may be interpreted instead as a Lebesgue integral over the set $(0, \infty)$. Since both of these kinds of integral agree, one is free to choose the first method to calculate the value of the integral, even if one ultimately wishes to regard it as a Lebesgue integral. Thus improper integrals are clearly useful tools for obtaining the actual values of integrals.

In other cases, however, a Lebesgue integral between finite endpoints may not even be defined, because the integrals of the positive and negative parts of f are both infinite, but the improper Riemann integral may still exist. Such cases are "properly improper" integrals, i.e. their values cannot be defined except as such limits. For example,

$$\int_0^\infty \frac{\sin(x)}{x}\mathrm{d}x$$

cannot be interpreted as a Lebesgue integral, since

$$\int_0^\infty \left|\frac{\sin(x)}{x}\right|\mathrm{d}x = \infty.$$

But $f(x) = \sin(x) / x$ is nevertheless Riemann integrable between any two finite endpoints, and its integral between 0 and " is usually understood as the limit of the Riemann integral:

$$\int_0^\infty \frac{\sin(x)}{x}\mathrm{d}x = \lim_{b\to\infty}\int_0^b \frac{\sin(x)}{x}\mathrm{d}x = \frac{\pi}{2}.$$

Singularities

One can speak of the *singularities* of an improper integral, meaning those points of the extended real number line at which limits are used.

Cauchy Principal Value

Consider the difference in values of two limits:

$$\lim_{a\to 0+}\left(\int_{-1}^{-a}\frac{\mathrm{d}x}{x}+\int_a^1\frac{\mathrm{d}x}{x}\right)=0,$$

$$\lim_{a\to 0+}\left(\int_{-1}^{-a}\frac{\mathrm{d}x}{x}+\int_{2a}^1\frac{\mathrm{d}x}{x}\right)=-\ln 2.$$

The former is the Cauchy principal value of the otherwise ill-defined expression

$$\int_{-1}^1\frac{\mathrm{d}x}{x}\ (\text{which gives } -\infty+\infty).$$

Similarly, we have

$$\lim_{a\to\infty}\int_{-a}^a\frac{2x\mathrm{d}x}{x^2+1}=0,$$

but

$$\lim_{a\to\infty}\int_{-2a}^a\frac{2x\mathrm{d}x}{x^2+1}=-\ln 4.$$

The former is the principal value of the otherwise ill-defined expression

$$\int_{-\infty}^\infty\frac{2x\mathrm{d}x}{x^2+1}\ (\text{which gives } -\infty+\infty).$$

All of the above limits are cases of the indeterminate form $\infty - \infty$.

These pathologies do not affect "Lebesgue-integrable" functions, that is, functions the integrals of whose absolute values are finite.

Summability

An indefinite integral may diverge in the sense that the limit defining it may not exist. In this case, there are more sophisticated definitions of the limit which can produce a convergent value for the improper integral. These are called summability methods.

One summability method, popular in Fourier analysis, is that of Cesàro summation. The integral

$$\int_0^\infty f(x)\mathrm{d}x$$

is Cesàro summable (C, α) if

$$\lim_{\lambda\to\infty}\int_0^\lambda\left(1-\frac{x}{\lambda}\right)^\alpha f(x)\mathrm{d}x$$

exists and is finite (Titchmarsh 1948, §1.15). The value of this limit, should it exist, is the (C, α) sum of the integral.

An integral is (C, 0) summable precisely when it exists as an improper integral. However, there are integrals which are (C, α) summable for $\alpha > 0$ which fail to converge as improper integrals (in the sense of Riemann or Lebesgue). One example is the integral

$$\int_0^\infty \sin x\,\mathrm{d}x$$

which fails to exist as an improper integral, but is (C,α) summable for every $\alpha > 0$. This is an integral version of Grandi's series.

Chapter 8

Line Integral

In mathematics, a line integral (sometimes called a path integral, contour integral, or curve integral; not to be confused with calculating arc length using integration) is an integral where the function to be integrated is evaluated along a curve.

The function to be integrated may be a scalar field or a vector field. The value of the line integral is the sum of values of the field at all points on the curve, weighted by some scalar function on the curve (commonly arc length or, for a vector field, the scalar product of the vector field with a differential vector in the curve).

This weighting distinguishes the line integral from simpler integrals defined on intervals. Many simple formulae in physics (for example, W=F.s) have natural continuous analogs in terms of line integrals ($W=\int_C \mathbf{F} \cdot d\mathbf{s}$). The line integral finds the work done on an object moving through an electric or gravitational field, for example.

Vector Calculus

In qualitative terms, a line integral in vector calculus can be thought of as a measure of the total effect of a given field along a given curve. More specifically, the line integral over a scalar field can be interpreted as the area under the field carved out by a particular curve. This can be visualized as the surface created by $z = f(x,y)$ and a curve C in the x-y plane. The line integral of f would be the area of the "curtain" created when the points of the surface that are directly over C are carved out.

Line Integral of a Scalar Field

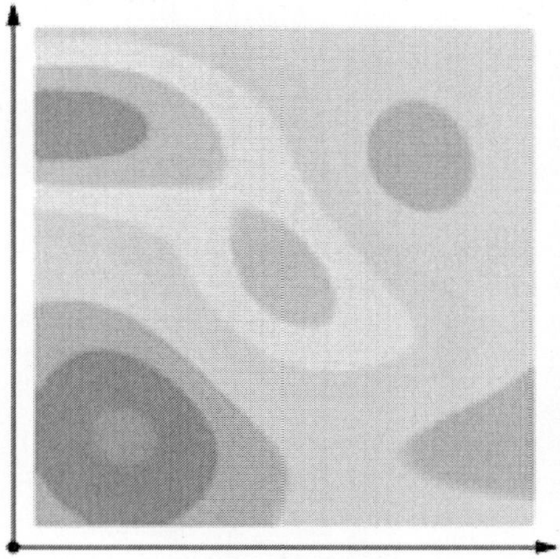

Figure: *The line integral over a scalar field* f *can be thought of as the area under the curve* C *along a surface* z = f(x,y)*, described by the field.*

Definition

For some scalar field $f: U \subseteq \mathrm{R}^n \to \mathrm{R}$, the line integral along a piecewise smooth curve $C \subset U$ is defined as

$$\int_C f\,ds = \int_a^b f(\mathrm{r}(t))\,|\,\mathrm{r}'(t)\,|dt.$$

where r: [a, b] $\to C$ is an arbitrary bijective parametrization of the curve C such that r(a) and r(b) give the endpoints of C and $a < b$. The function f is called the integrand, the curve C is the domain of integration, and the symbol ds may be intuitively interpreted as an elementary arc length. Line integrals of scalar fields over a curve C do not depend on the chosen parametrization r of C.

Geometrically, when the scalar field f is defined over a plane (n=2), its graph is a surface $z=f(x,y)$ in space, and the line integral gives the (signed) cross-sectional area bounded by the curve C and the graph of f.

Derivation

For a line integral over a scalar field, the integral can be constructed from a Riemann sum using the above definitions of f, C and a parametrization r of C. This can be done by partitioning the interval $[a,b]$ into n sub-intervals $[t_{i-1}, t_i]$ of length $\Delta t = (b - a)/n$, then $r(t_i)$ denotes some point, call it a sample point, on the curve C. We can use the set of sample points $\{r(t_i) : 1 \leq i \leq n\}$ to approximate the curve C by a polygonal path by introducing a straight line piece between each of the sample points $r(t_{i-1})$ and $r(t_i)$. We then label the distance between each of the sample points on the curve as Δs_i. The product of $f(r(t_i))$ and Δs_i can be associated with the signed area of a rectangle with a height and width of $f(r(t_i))$ and Δs_i respectively. Taking the limit of the sum of the terms as the length of the partitions approaches zero gives us I

$$I = \lim_{\Delta s_i \to 0} \sum_{i=1}^{n} f(r(t_i))\Delta s_i.$$

We note that, by the mean value theorem, the distance between subsequent points on the curve, is

$$\Delta s_i = | r(t_i + \Delta t) - r(t_i) | \approx | r'(t_i) | \Delta t.$$

Substituting this in the above Riemann sum yields

$$I = \lim_{\Delta t \to 0} \sum_{i=1}^{n} f(r(t_i)) \, | r'(t_i) | \, \Delta t$$

which is the Riemann sum for the integral

$$I = \int_a^b f(r(t)) \, | r'(t) | dt.$$

Line Integral of a Vector Field

Definition: For a vector field $F : U \subseteq R^n \to R^n$, the line integral along a piecewise smooth curve $C \subset U$, in the direction of r, is defined as

$$\int_C F(r) \cdot dr = \int_a^b F(r(t)) \cdot r'(t) dt.$$

where $\cdot$ is the dot product and r: [a, b] $\to$ C is a bijective parametrization of the curve C such that $r(a)$ and $r(b)$ give the endpoints of C.

A line integral of a scalar field is thus a line integral of a vector field where the vectors are always tangential to the line.

Line integrals of vector fields are independent of the parametrization r in absolute value, but they do depend on its orientation. Specifically, a reversal in the orientation of the parametrization changes the sign of the line integral.

Derivation

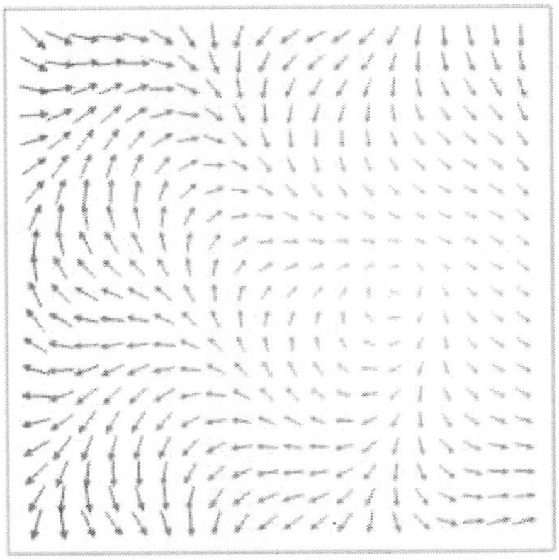

***Figure:** The trajectory of a particle (in red) along a curve inside a vector field. Starting from a, the particle traces the path C along the vector field F. The dot product (green line) of its displacement vector (red arrow) and the field vector (blue arrow) defines an area under a curve, which is equivalent to the path's line integral. (Click on image for a detailed description.)*

The line integral of a vector field can be derived in a manner very similar to the case of a scalar field. Again using the above definitions of F, C and its parametrization r(t), we construct the integral from a Riemann sum.

Partition the interval $[a,b]$ into n intervals of length $\Delta t = (b - a)/n$. Letting t_i be the ith point on $[a,b]$, then $r(t_i)$ gives us the position of the ith point on the curve. However, instead of calculating up the distances between subsequent points, we need to calculate their displacement vectors, Δr_i. As before, evaluating F at all the points on the curve and taking the dot product with each displacement vector gives us the infinitesimal contribution of each partition of F on C. Letting the size of the partitions go to zero gives us a sum

$$I = \lim_{\Delta t \to 0} \sum_{i=1}^{n} F(r(t_i)) \cdot \Delta r_i$$

By the mean value theorem, we see that the displacement vector between adjacent points on the curve is

$$\Delta \mathrm{r}_i = \mathrm{r}(t_i + \Delta t) - \mathrm{r}(t_i) \approx \mathrm{r}'(t_i)\Delta t$$

Substituting this in the above Riemann sum yields

$$I = \lim_{\Delta t \to 0} \sum_{i=1}^{n} \mathrm{F}(\mathrm{r}(t_i)) \cdot \mathrm{r}'(t_i)\Delta t$$

which is the Riemann sum for the integral defined above.

Path Independence

If a vector field F is the gradient of a scalar field G (i.e. if F is conservative), that is,

$$\nabla G = \mathrm{F},$$

then the derivative of the composition of G and $\mathrm{r}(t)$ is

$$\frac{dG(\mathrm{r}(t))}{dt} = \nabla G(\mathrm{r}(t)) \cdot \mathrm{r}'(t) = \mathrm{F}(\mathrm{r}(t)) \cdot \mathrm{r}'(t)$$

which happens to be the integrand for the line integral of F on $\mathrm{r}(t)$. It follows that, given a path C, then

$$\int_C \mathrm{F}(\mathrm{r}) \cdot d\mathrm{r} = \int_a^b \mathrm{F}(\mathrm{r}(t)) \cdot \mathrm{r}'(t)dt = \int_a^b \frac{dG(\mathrm{r}(t))}{dt}dt = G(\mathrm{r}(b)) - G(\mathrm{r}(a)).$$

In other words, the integral of F over C depends solely on the values of G in the points $\mathrm{r}(b)$ and $\mathrm{r}(a)$ and is thus independent of the path between them.

For this reason, a line integral of a conservative vector field is called *path independent.*

Applications

The line integral has many uses in physics. For example, the work done on a particle travelling on a curve C inside a force field represented as a vector field F is the line integral of F on C.

Complex Line Integral

The line integral is a fundamental tool in complex analysis. Suppose U is an open subset of the complex plane C, $f: U \to$ C is a function, and $L \subset U$ is a rectifiable curve parametrized by $\gamma : [a, b] \to L$, $\gamma(t)=x(t)+iy(t)$. The line integral

$$\int_L f(z)dz = \int_\gamma f(z)dz$$

may be defined by subdividing the interval $[a, b]$ into $a = t_0 < t_1 < \ldots < t_n = b$ and considering the expression

$$\sum_{k=1}^{n} f(\gamma(t_k))[\gamma(t_k) - \gamma(t_{k-1})] = \sum_{k=1}^{n} f(\gamma_k)\Delta\gamma_k = S_n.$$

The integral is then the limit of this Riemann sum as the lengths of the subdivision intervals approach zero. If the parametrization γ is continuously differentiable, the line integral can be evaluated as an integral of a function of a real variable:

$$\int_L f(z)dz = \int_a^b f(y))\, y'(t)\, dt$$

When $L$is a closed curve, that is, its initial and final points coincide, the notation

$$\oint_L f(z)dz$$

is often used for the line integral of f along L. A closed curve line integral is sometimes referred to as a cyclic integral in engineering applications.

The line integrals of complex functions can be evaluated using a number of techniques: the integral may be split into real and imaginary parts reducing the problem to that of evaluating two real-valued line integrals, the Cauchy integral formula may be used in other circumstances. If the line integral is a closed curve in a region where the function is analytic and containing no singularities, then the value of the integral is simply zero; this is a consequence of the Cauchy integral theorem. The residue theorem allows contour integrals to be used in the complex plane to find integrals of real-valued functions of a real variable.

Example

Consider the function $f(z)=1/z$, and let the contour L be the unit circle about 0, parametrized by $\gamma(t)=e^{it}$ with t in $[0, 2\pi]$ (which generates the circle counterclockwise). Substituting, we find

$$\begin{aligned}\oint_L f(z)dz &= \int_0^{2\pi} \frac{1}{e^{it}} ie^{it}\, dt = i\int_0^{2\pi} e^{-it}e^{it}\, dt \\ &= i\int_0^{2\pi} dt = i(2\pi - 0) = 2\pi i.\end{aligned}$$

Here we have used the fact that any complex number z can be written as re^{it} where r is the modulus of z. On the unit circle this is fixed to 1, so the only variable left is the angle, which is denoted by t.

This answer can be also verified by the Cauchy integral formula.

Relation between the Line Integral of a Vector Field and the Complex Line Integral

Viewing complex numbers as 2-dimensional vectors, the line integral of a 2-dimensional vector field corresponds to the real part of the line integral of the conjugate of the corresponding complex function of a complex variable. More specifically, if $\mathrm{r}(t) = x(t)\mathrm{i} + y(t)\mathrm{j}$ and $f(z) = u(z) + iv(z)$, then:

$$\int_L \overline{f(z)}dz = \int_L (u - iv)dz = \int_L (u\mathrm{i} + v\mathrm{j})\cdot dr - i\int_L (v\mathrm{i} - u\mathrm{j})\cdot dr,$$

provided that both integrals on the right hand side exist, and that the parametrization γ of L has the same orientation as r (just expand the Riemann sum for the lefthand integral and take the limit). Due to the Cauchy-Riemann equations the curl of the vector field corresponding to the conjugate of a holomorphic function is zero. This relates through Stokes' theorem both types of line integral being zero.

Quantum Mechanics

The "path integral formulation" of quantum mechanics actually refers not to path integrals in this sense but to functional integrals, that is, integrals over a space of paths, of a function *of* a possible path.

However, path integrals in the sense of this article are important in quantum mechanics; for example, complex contour integration is often used in evaluating probability amplitudes in quantum scattering theory.

Riemann Sum

In mathematics, a Riemann sum is a summation of a large number of small partitions of a region. It may be used to define the integration operation. The method was named after German mathematician Bernhard Riemann.

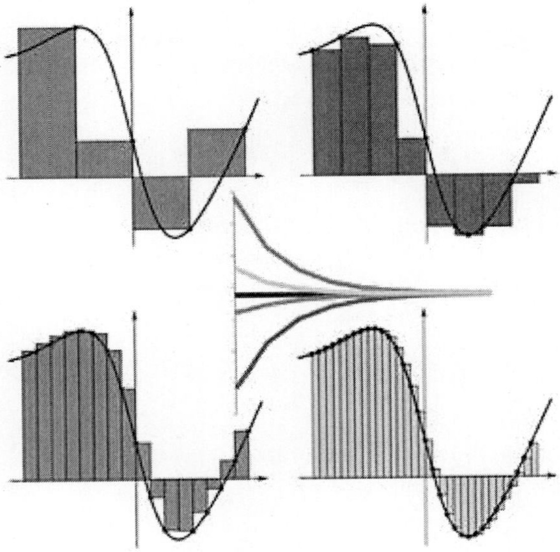

Figure: *Four of the Riemann summation methods for approximating the area under curves. Right and left methods make the approximation using the right and left endpoints of each subinterval, respectively. Maximum and minimum methods make the approximation using the largest and smallest endpoint values of each subinterval, respectively. The values of the sums converge as the subintervals halve from top-left to bottom-right.*

Methods

The four methods of Riemann summation are usually best approached with partitions of equal size. The interval $[a, b]$ is therefore divided into n subintervals, each of length

$$\Delta x = \frac{b-a}{n}.$$

The points in the partition will then be

$$a, a+\Delta x, a+2\Delta x, \ldots, a+(n-2)\Delta x, a+(n-1)\Delta x, b.$$

Left Sum

For the left Riemann sum, approximating the function by its value at the left-end point gives multiple rectangles with base Δx and height $f(a + i\Delta x)$. Doing this for $i = 0, 1, \ldots, n-1$, and adding up the resulting areas gives

$$\Delta x\left[f(a)+f(a+\Delta x)+f(a+2\Delta x)+\cdots+f(b-\Delta x)\right].$$

The left Riemann sum amounts to an n if *f* is monotonically decreasing on this interval, and an underestimation if it is monotonically increasing.

Right Sum

f is here approximated by the value at the right endpoint. This gives multiple rectangles with base Δx and height $f(a + i\Delta x)$. Doing this for $i = 1,..., n$, and adding up the resulting areas produces

$$\Delta x\left[f(a+\Delta x)+f(a+2\Delta x)+\cdots+f(b)\right].$$

The right Riemann sum amounts to an underestimation if *f* is monotonically decreasing, and an overestimation if it is monotonically increasing.

Middle Sum

Approximating *f* at the midpoint of intervals gives $f(a + Q/2)$ for the first interval, for the next one $f(a + 3Q/2)$, and so on until $f(b - Q/2)$. Summing up the areas gives

$$Q\left[f(a+\tfrac{Q}{2})+f(a+\tfrac{3Q}{2})+\cdots+f(b-\tfrac{Q}{2})\right].$$

The error of this formula will be

$$\left|\int_a^b f(x)dx - A_{\text{mid}}\right| \le \frac{M_2(b-a)^3}{24n^2},$$

where M_2 is the maximum value of the absolute value of $f''(x)$ on the interval.

Trapezoidal Rule

In this case, the values of the function *f* on an interval are approximated by the average of the values at the left and right endpoints. In the same manner as above, a simple calculation using the area formula

$$A = \tfrac{1}{2}h(b_1 + b_2)$$

for a trapezium with parallel sides b_1, b_2 and height h produces

$$\tfrac{1}{2}Q\left[f(a)+2f(a+Q)+2f(a+2Q)+2f(a+3Q)+\cdots+f(b)\right].$$

The error of this formula will be

$$\left|\int_a^b f(x)dx - A_{\text{trap}}\right| \le \frac{M_2(b-a)^3}{12n^2},$$

where M_2 is the maximum value of the absolute value of $f''(x)$.

The approximation obtained with the trapezoid rule for a function is the same as the average of the left hand and right hand sums of that function.

Example

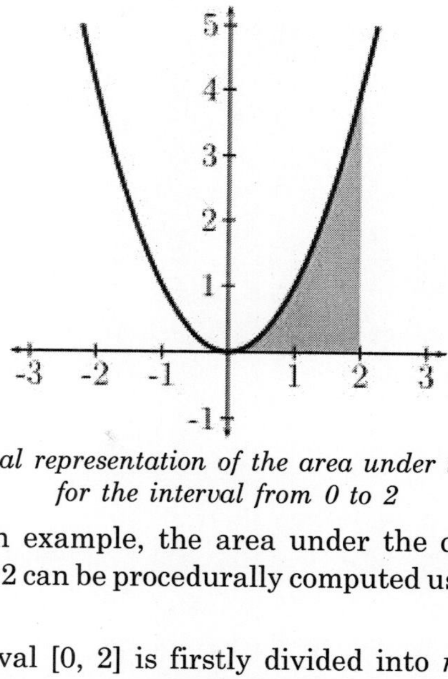

Figure: *A visual representation of the area under the curve* y = x². *for the interval from 0 to 2*

Taking an example, the area under the curve of $y = x^2$ between 0 and 2 can be procedurally computed using Riemann's method.

The interval [0, 2] is firstly divided into n subintervals, each of which is given a width of $\frac{2}{n}$; these are the widths of the Riemann rectangles.

Because the right Riemann sum is to be used, the sequence of x coordinates for the boxes will be $x_1, x_2, \ldots, x_n$. Therefore, the sequence of the heights of the boxes will be $x_1^2, x_2^2, \ldots, x_n^2$. It is an important fact that $x_i = \frac{2i}{n}$, and $x_n = 2$.

The area of each box will be $\frac{2}{n} \times x_i^2$ and therefore the nth right Riemann sum will be:

$$
\begin{aligned}
S &= \frac{2}{n}\times\left(\frac{2}{n}\right)^2 + \cdots + \frac{2}{n}\times\left(\frac{2i}{n}\right)^2 + \cdots + \frac{2}{n}\times\left(\frac{2n}{n}\right)^2 \\
&= \frac{8}{n^3}\left(1+\cdots+i^2+\cdots+n^2\right) \\
&= \frac{8}{n^3}\left(\frac{n(n+1)(2n+1)}{6}\right) \\
&= \frac{8}{n^3}\left(\frac{2n^3+3n^2+n}{6}\right) \\
&= \frac{8}{3}+\frac{4}{n}+\frac{4}{3n^2}
\end{aligned}
$$

If the limit is viewed as $n \to \infty$, it can be concluded that the approximation approaches the actual value of the area under the curve as the number of boxes increases. Hence:

$$\lim_{n\to\infty} S = \lim_{n\to\infty}\left(\frac{8}{3}+\frac{4}{n}+\frac{4}{3n^2}\right) = \frac{8}{3}$$

This method agrees with the definite integral as calculated in more mechanical ways:

$$\int_0^2 x^2\,dx = \frac{8}{3}$$

Riemann–Stieltjes Integral

In mathematics, the Riemann–Stieltjes integral is a generalization of the Riemann integral, named after Bernhard Riemann and Thomas Joannes Stieltjes. The definition of this integral was first published in 1894 by Stieltjes. It serves as an instructive and useful precursor of the Lebesgue integral.

Definition

The Riemann–Stieltjes integral of a real-valued function f of a real variable with respect to a real function g is denoted by

$$\int_a^b f(x)\,dg(x)$$

and defined to be the limit, as the mesh of the partition

$$P = \{a = x_0 < x_1 < \cdots < x_n = b\}$$

of the interval $[a, b]$ approaches zero, of the approximating sum

$$S(P,f,g)=\sum_{i=0}^{n-1} f(c_i)(g(x_{i+1})-g(x_i))$$

where c_i is in the i-th subinterval $[x_i, x_{i+1}]$. The two functions f and g are respectively called the integrand and the integrator.

The "limit" is here understood to be a number A (the value of the Riemann–Stieltjes integral) such that for every $\varepsilon > 0$, there exists $\delta > 0$ such that for every partition P with mesh(P) < δ, and for every choice of points c_i in $[x_i, x_{i+1}]$,

$$|S(P,f,g)-A|<\varepsilon.$$

Generalized Riemann–Stieltjes Integral

A slight generalization, introduced by Pollard (1920) and now standard in analysis, is to consider in the above definition partitions P that *refine* another partition P_ε, meaning that P arises from P_ε by the addition of points, rather than from partitions with a finer mesh. Specifically, the generalized Riemann–Stieltjes integral of f with respect to g is a number A such that for every $\varepsilon > 0$ there exists a partition P_ε such that for every partition P that refines P_ε,

$$|S(P,f,g)-A|<\varepsilon$$

for every choice of points c_i in $[x_i, x_{i+1}]$.

This generalization exhibits the Riemann–Stieltjes integral as the Moore–Smith limit on the directed set of partitions of $[a, b]$ (McShane 1952). Hildebrandt (1938) calls it the Pollard–Moore–Stieltjes integral.

Darboux Sums

The Riemann–Stieltjes integral can be efficiently handled using an appropriate generalization of Darboux sums. For a partition P and a nondecreasing function g on $[a, b]$ define the upper Darboux sum of f with respect to g by

$$U(P,f,g)=\sum_{i=1}^{n} \sup_{x\in[x_i,x_{i+1}]} f(x)(g(x_{i+1})-g(x_i))$$

and the lower sum by

$$L(P,f,g)=\sum_{i=1}^{n} \inf_{x\in[x_i,x_{i+1}]} f(x)(g(x_{i+1})-g(x_i)).$$

Then the generalized Riemann–Stieltjes of f with respect to g exists if and only if, for every $\varepsilon > 0$, there exists a partition P such that

$$U(P,f,g) - L(P,f,g) < \varepsilon.$$

Furthermore, f is Riemann–Stieltjes integrable with respect to g (in the classical sense) if

$$\lim_{\text{mesh}(P)\to 0}[U(P,f,g) - L(P,f,g)] = 0.$$

Properties and Relation to the Riemann Integral

If g should happen to be everywhere differentiable, then the Riemann–Stieltjes integral may still be different from the Riemann integral of $f(x)g'(x)$ given by

$$\int_a^b f(x)g'(x)dx,$$

for example, if the derivative is unbounded. But if the derivative is continuous, they will be the same. This condition is also satisfied if g is the (Lebesgue) integral of its derivative; in this case g is said to be absolutely continuous.

However, g may have jump discontinuities, or may have derivative zero *almost* everywhere while still being continuous and increasing (for example, g could be the Cantor function), in either of which cases the Riemann–Stieltjes integral is not captured by any expression involving derivatives of g.

The Riemann–Stieltjes integral admits integration by parts in the form

$$\int_a^b f(x)dg(x) = f(b)g(b) - f(a)g(a) - \int_a^b g(x)df(x).$$

and the existence of either integral implies the existence of the other (Hille & Phillips 1974, §3.3).

Existence of the Integral

The best simple existence theorem states that if f is continuous and g is of bounded variation on $[a, b]$, then the integral exists. A function g is of bounded variation if and only if it is the difference between two monotone functions. If g is not of bounded variation, then there will be continuous functions which cannot be integrated with respect to g. In general, the integral is not well-defined if f and g share any points of discontinuity, but this sufficient condition is not necessary.

On the other hand, a classical result of Young (1936) states that the integral is well-defined if f is α-Hölder continuous and g is β-Hölder continuous with $\alpha + \beta > 1$.

Application to Probability Theory

If g is the cumulative probability distribution function of a random variable X that has a probability density function with respect to Lebesgue measure, and f is any function for which the expected value E(|$f(X)$|) is finite, then the probability density function of X is the derivative of g and we have

$$E(f(X)) = \int_{-\infty}^{\infty} f(x)g'(x)dx.$$

But this formula does not work if X does not have a probability density function with respect to Lebesgue measure. In particular, it does not work if the distribution of X is discrete (i.e., all of the probability is accounted for by point-masses), and even if the cumulative distribution function g is continuous, it does not work if g fails to be absolutely continuous (again, the Cantor function may serve as an example of this failure). But the identity

$$E(f(X)) = \int_{-\infty}^{\infty} f(x)dg(x)$$

holds if g is *any* cumulative probability distribution function on the real line, no matter how ill-behaved. In particular, no matter how ill-behaved the cumulative distribution function g of a random variable X, if the moment E(X^n) exists, then it is equal to

$$E(X^n) = \int_{-\infty}^{\infty} x^n \, dg(x).$$

Application to Functional Analysis

The Riemann–Stieltjes integral appears in the original formulation of F. Riesz's theorem which represents the dual space of the Banach space $C[a,b]$ of continuous functions in an interval $[a,b]$ as Riemann–Stieltjes integrals against functions of bounded variation. Later, that theorem was reformulated in terms of measures.

The Riemann–Stieltjes integral also appears in the formulation of the spectral theorem for (non-compact) self-adjoint (or more generally, normal) operators in a Hilbert space.

In this theorem, the integral is considered with respect to a spectral family of projections.

Generalization

An important generalization is the Lebesgue–Stieltjes integral which generalizes the Riemann–Stieltjes integral in a way analogous to how the Lebesgue integral generalizes the Riemann integral. If improper Riemann–Stieltjes integrals are allowed, the Lebesgue integral is not strictly more general than the Riemann–Stieltjes integral.

The Riemann–Stieltjes integral also generalizes to the case when either the integrand f or the integrator g take values in a Banach space. If $g : [a,b] \to X$ takes values in the Banach space X, then it is natural to assume that it is of strongly bounded variation, meaning that

$$\sup \sum_i \| g(t_i) - g(t_{i+1}) \|_X < \infty$$

the supremum being taken over all finite partitions

$$a = t_0 \le t_1 \le \cdots \le t_n = b$$

of the interval $[a,b]$. This generalization plays a role in the study of semigroups, via the Laplace–Stieltjes transform.

Real Number

In mathematics, a real number is a value that represents a quantity along a continuous line. The real numbers include all the rational numbers, such as the integer "5 and the fraction 4/3, and all the irrational numbers such as "2 (1.41421356... the square root of two, an irrational algebraic number) and π (3.14159265..., a transcendental number). Real numbers can be thought of as points on an infinitely long line called the number line or real line, where the points corresponding to integers are equally spaced. Any real number can be determined by a possibly infinite decimal representation such as that of 8.632, where each consecutive digit is measured in units one tenth the size of the previous one. The real line can be thought of as a part of the complex plane, and correspondingly, complex numbers include real numbers as a special case.

These descriptions of the real numbers are not sufficiently rigorous by the modern standards of pure mathematics. The discovery of a suitably rigorous definition of the real numbers — indeed, the realization that a better definition was needed — was one of the most important developments of 19th century mathematics. The currently standard axiomatic definition is that real numbers form the unique Archimedean complete totally ordered field (R,+,·,<), up to isomorphism, whereas popular constructive definitions of real numbers include declaring them as equivalence classes of Cauchy sequences of rational numbers, Dedekind cuts, or certain infinite "decimal representations", together with precise interpretations for the arithmetic operations and the order relation. These definitions are equivalent in the realm of classical mathematics.

The reals are uncountable, that is, while both the set of all natural numbers and the set of all real numbers are infinite sets, there can be no one-to-one function from the real numbers to the natural numbers: the cardinality of the set of all real numbers (denoted $\mathfrak{c}$ and called cardinality of the continuum) is strictly greater than the cardinality of the set of all natural numbers (denoted $\aleph_0$). The statement that there is no subset of the reals with cardinality strictly greater than $\mathfrak{c}$ and strictly smaller than is known as the continuum hypothesis. It is known to be neither provable nor refutable using the axioms of Zermelo–Fraenkel set theory $\aleph_0$, the standard foundation of modern mathematics, provided ZF set theory is consistent.

Basic Properties

A real number may be either rational or irrational; either algebraic or transcendental; and either positive, negative, or zero. Real numbers are used to measure continuous quantities. They may be expressed by decimal representations that have an infinite sequence of digits to the right of the decimal point; these are often represented in the same form as 324.823122147... The ellipsis (three dots) indicate that there would still be more digits to come.

More formally, real numbers have the two basic properties of being an ordered field, and having the least upper bound property. The first says that real numbers comprise a field, with addition and multiplication as well as division by nonzero numbers, which can be totally ordered on a number line in a way compatible with addition and multiplication. The second says that if a nonempty set of real numbers has an upper bound, then it has a real least upper bound. The second condition distinguishes the real numbers from the rational numbers: for example, the set of rational numbers whose square is less than 2 is a set with an upper bound (e.g. 1.5) but no (rational) least upper bound: hence the rational numbers do not satisfy the least upper bound property.

In Physics?

In the physical sciences, most physical constants such as the universal gravitational constant, and physical variables, such as position, mass, speed, and electric charge, are modelled using real numbers. In fact, the fundamental physical theories such as classical mechanics, electromagnetism, quantum mechanics, general relativity and the standard model are described using mathematical structures, typically smooth manifolds or Hilbert spaces, that are based on the real numbers although actual measurements of physical quantities are of finite accuracy and precision.

In some recent developments of theoretical physics stemming from the holographic principle, the Universe is seen fundamentally as an information store, essentially zeroes and ones, organized in much less geometrical fashion and manifesting itself as space-time and particle fields only on a more superficial level. This approach removes the real number system from its foundational role in physics and even prohibits the existence of infinite precision real numbers in the physical universe by considerations based on the Bekenstein bound.

In Computation

Computer arithmetic cannot directly operate on real numbers, but only on a finite subset of rational numbers, limited by the number of bits used to store them, whether as floating-point numbers or arbitrary precision numbers. However, computer algebra systems can operate on irrational quantities exactly by manipulating formulas for them (such as $\sqrt{2}$, $\arcsin\left(\frac{2}{23}\right)$, or $\int_0^1 x^x\,dx$) rather than their rational or decimal approximation; however, it is not in general possible to determine whether two such expressions are equal (the constant problem).

A real number is called *computable* if there exists an algorithm that yields its digits. Because there are only countably many algorithms, but an uncountable number of reals, almost all real numbers fail to be computable. Moreover, the equality of two computable numbers is an undecidable problem. Some constructivists accept the existence of only those reals that are computable. The set of definable numbers is broader, but still only countable

Notation

Mathematicians use the symbol R (or alternatively, $\mathbb{R}$, the letter "R" in blackboard bold, Unicode $\mathbb{R}$ – U+211D) to represent the set of all real numbers (as this set is naturally endowed with a structure of field, the expression *field of the real numbers* is more frequently used than *set of all real numbers*). The notation R^n refers to the Cartesian product of n copies of R, which is an n-dimensional vector space over the field of the real numbers; this vector space may be identified to the n-dimensional space of Euclidean geometry as soon as a coordinate system has been chosen in the latter. For example, a value from R^3 consists of three real numbers and specifies the coordinates of a point in 3-dimensional space.

In mathematics, *real* is used as an adjective, meaning that the underlying field is the field of the real numbers (or *the real field*). For example *real matrix, real polynomial* and *real Lie algebra*. As a substantive, the term is used almost strictly in reference to the real numbers themselves (e.g., The "set of all reals").

History

Simple fractions have been used by the Egyptians around 1000 BC; the Vedic "Sulba Sutras" ("The rules of chords") in, ca. 600 BC, include what may be the first 'use' of irrational numbers. The concept of irrationality was implicitly accepted by early Indian mathematicians since Manava (c. 750–690 BC), who were aware that the square roots of certain numbers such as 2 and 61 could not be exactly determined. Around 500 BC, the Greek mathematicians led by Pythagoras realized the need for irrational numbers, in particular the irrationality of the square root of 2.

The Middle Ages brought the acceptance of zero, negative, integral, and fractional numbers, first by Indian and Chinese mathematicians, and then by Arabic mathematicians, who were also the first to treat irrational numbers as algebraic objects, which was made possible by the development of algebra. Arabic mathematicians merged the concepts of "number" and "magnitude" into a more general idea of real numbers. The Egyptian mathematician Abû Kâmil Shujâ ibn Aslam (c. 850–930) was the first to accept irrational numbers as solutions to quadratic equations or as coefficients in an equation, often in the form of square roots, cube roots and fourth roots.

In the 16th century, Simon Stevin created the basis for modern decimal notation, and insisted that there is no difference between rational and irrational numbers in this regard.

In the 17th century, Descartes introduced the term "real" to describe roots of a polynomial, distinguishing them from "imaginary" ones.

In the 18th and 19th centuries there was much work on irrational and transcendental numbers. Johann Heinrich Lambert (1761) gave the first flawed proof that π cannot be rational; Adrien-Marie Legendre (1794) completed the proof, and showed that π is not the square root of a rational number. Paolo Ruffini (1799) and Niels Henrik Abel (1842) both constructed proofs of Abel–Ruffini theorem: that the general quintic or higher equations cannot be solved by a general formula involving only arithmetical operations and roots.

Évariste Galois (1832) developed techniques for determining whether a given equation could be solved by radicals, which gave rise to the field of Galois theory. Joseph Liouville (1840) showed that neither e nor e^2 can be a root of an integer quadratic equation, and then established existence of transcendental numbers, the proof being subsequently displaced by Georg Cantor (1873). Charles Hermite (1873) first proved that e is transcendental, and Ferdinand von Lindemann (1882), showed that π is transcendental. Lindemann's proof was much simplified by Weierstrass (1885), still further by David Hilbert (1893), and has finally been made elementary by Adolf Hurwitz and Paul Gordan.

The development of calculus in the 18th century used the entire set of real numbers without having defined them cleanly. The first rigorous definition was given by Georg Cantor in 1871. In 1874 he showed that the set of all real numbers is uncountably infinite but the set of all algebraic numbers is countably infinite. Contrary to widely held beliefs, his first method was not his famous diagonal argument, which he published in 1891.

Construction of the Real Numbers

In mathematics, there are several ways of defining the real number system as an ordered field. The *synthetic* approach gives a list of axioms for the real numbers as a *complete ordered field.* Under the usual axioms of set theory, one can show that these axioms are categorical, in the sense that there is a model for the axioms, and any two such models are isomorphic. Any one of these models must be explicitly constructed, and most of these models are built using the basic properties of the rational number system as an ordered field.

Synthetic Approach

The synthetic approach axiomatically defines the real number system as a complete ordered field. Precisely, this means the following. A *model for the real number system* consists of a set R, two distinct elements 0 and 1 of R, two binary operations + and * on R (called *addition* and *multiplication*, resp.), a binary relation ≤ on R, satisfying the following properties.

1. (R, +, *) forms a field. In other words,
   - For all x, y, and z in R, $x + (y + z) = (x + y) + z$ and $x * (y * z) = (x * y) * z$. (associativity of addition and multiplication)
   - For all x and y in R, $x + y = y + x$ and $x * y = y * x$. (commutativity of addition and multiplication)
   - For all x, y, and z in R, $x * (y + z) = (x * y) + (x * z)$. (distributivity of multiplication over addition)
   - For all x in R, $x + 0 = x$. (existence of additive identity)

— 0 is not equal to 1, and for all x in R, $x * 1 = x$. (existence of multiplicative identity)

— For every x in R, there exists an element "x in R, such that $x + (-x) = 0$. (existence of additive inverses)

— For every $x \neq 0$ in R, there exists an element x^{-1} in R, such that $x * x^{-1} = 1$. (existence of multiplicative inverses)

2. (R, ≤) forms a totally ordered set. In other words,

   — For all x in R, $x \leq x$. (reflexivity)

   — For all x and y in R, if $x \leq y$ and $y \leq x$, then $x = y$. (antisymmetry)

   — For all x, y, and z in R, if $x \leq y$ and $y \leq z$, then $x \leq z$. (transitivity)

   — For all x and y in R, $x \leq y$ or $y \leq x$. (totalness)

3. The field operations + and * on R are compatible with the order ≤. In other words,

   — For all x, y and z in R, if $x \leq y$, then $x + z \leq y + z$. (preservation of order under addition)

   — For all x and y in R, if $0 \leq x$ and $0 \leq y$, then $0 \leq x * y$ (preservation of order under multiplication)

4. The order ≤ is *complete* in the following sense: every non-empty subset of R bounded above has a least upper bound. In other words,

   — If A is a non-empty subset of R, and if A has an upper bound, then A has a least upper bound u, such that for every upper bound v of A, $u \leq v$.

The final axiom, defining the order as Dedekind-complete, is most crucial. Without this axiom, we simply have the axioms which define a totally ordered field, and there are many non-isomorphic models which satisfy these axioms. This axiom implies that the Archimedean property applies for this field. Therefore, when the completeness axiom is added, it can be proved that any two models must be isomorphic, and so in this sense, there is only one complete ordered Archimedean field.

When we say that any two models of the above axioms are isomorphic, we mean that for any two models $(R, 0_R, 1_R, +_R, *_R,$

$\leq_R$) and (S, 0_S, 1_S, $+_S$, $*_S$, $\leq_S$), there is a bijection $f: R \to S$ preserving both the field operations and the order. Explicitly,

- f is both injective and surjective.
- $f(0_R) = 0_S$ and $f(1_R) = 1_S$.
- For all x and y in R, $f(x +_R y) = f(x) +_S f(y)$ and $f(x *_R y) = f(x) *_S f(y)$.
- For all x and y in R, $x \leq_R y$ if and only if $f(x) \leq_S f(y)$.

Explicit Constructions of Models

We shall not prove that any models of the axioms are isomorphic. Such a proof can be found in any number of modern analysis or set theory textbooks. We will sketch the basic definitions and properties of a number of constructions, however, because each of these is important for both mathematical and historical reasons. The first three, due to Georg Cantor/Charles Méray, Richard Dedekind and Karl Weierstrass/Otto Stolz all occurred within a few years of each other. Each has advantages and disadvantages. A major motivation in all three cases was the instruction of mathematics students.

Construction from Cauchy Sequences

If we have a space where Cauchy sequences are meaningful (such as a 'rational' metric space, i.e., a space in which distance is defined and takes rational values, or more generally a uniform space), a standard procedure to force all Cauchy sequences to converge is adding new points to the space (a process called completion). By starting with rational numbers and the metric $d(x,y) = |x - y|$, we can construct the real numbers, as will be detailed below. (A different metric on the rationals could result in the p-adic numbers instead.)

Let R be the set of Cauchy sequences of rational numbers. That is, sequences $x_1, x_2, x_3, \ldots$ of rational numbers such that for every rational $\varepsilon > 0$, there exists an integer N such that for all natural numbers $m,n > N$, $|x_m - x_n| < \varepsilon$. Here the vertical bars denote the absolute value.

Cauchy sequences (x) and (y) can be added and multiplied as follows:

$$(x_n) + (y_n) = (x_n + y_n)$$
$$(x_n) \times (y_n) = (x_n \times y_n).$$

Two Cauchy sequences are called *equivalent* if and only if the difference between them tends to zero.

Comparison between two cauchy sequences is possible as such : $(x_n) \geq (y_n)$ if and only if x is equivalent to y or there exists an integer N such that $x_n \geq y_n$ for all $n > N$.

This does indeed define an equivalence relation, it is compatible with the operations defined above, and the set R of all equivalence classes can be shown to satisfy all the usual axioms of the real numbers. This is remarkable because not all of these axioms necessarily apply to the rational numbers, which are being used to construct the sequences themselves. We can embed the rational numbers into the reals by identifying the rational number r with the equivalence class of the sequence $(r,r,r, \ldots)$.

The only real number axiom that does not follow easily from the definitions is the completeness of $\leq$, i.e. the least upper bound property. It can be proved as follows: Let S be a non-empty subset of R and U be an upper bound for S. Substituting a larger value if necessary, we may assume U is rational. Since S is non-empty, there is a rational number L such that $L < s$ for some s in S. Now define sequences of rationals (u_n) and (l_n) as follows:

Set $u_0 = U$ and $l_0 = L$.

For each n consider the number:

$m_n = (u_n + l_n)/2$

If m_n is an upper bound for S set:

$u_{n+1} = m_n$ and $l_{n+1} = l_n$

Otherwise set:

$l_{n+1} = m_n$ and $u_{n+1} = u_n$

This obviously defines two Cauchy sequences of rationals, and so we have real numbers $l = (l_n)$ and $u = (u_n)$. It is easy to prove, by induction on n that:

u_n is an upper bound for S for all n

and:

l_n is never an upper bound for S for any n

Thus u is an upper bound for S. To see that it is a least upper bound, notice that the limit of $(u_n - l_n)$ is 0, and so $l =$

u. Now suppose $b < u = l$ is a smaller upper bound for S. Since (l_n) is monotonic increasing it is easy to see that $b < l_n$ for some n. But l_n is not an upper bound for S and so neither is b. Hence u is a least upper bound for S and $\leq$ is complete.

A practical and concrete representative for an equivalence class representing a real number is provided by the representation to base b – in practice, b is usually 2 (binary), 8 (octal), 10 (decimal) or 16 (hexadecimal). For example, the number π = 3.14159... corresponds to the Cauchy sequence (3,3.1,3.14,3.141,3.1415,...). Notice that the sequence (0,0.9,0.99,0.999,0.9999,...) is equivalent to the sequence (1,1.0,1.00,1.000,1.0000,...); this shows that 0.999... = 1.

An advantage of this approach is that it does not use the linear order of the rationals, only the metric. Hence it generalizes to other metric spaces.

Stevin's Construction

It has been known since Simon Stevin that real numbers can be represented by decimals. We can take the infinite decimal expansion to be the definition of a real number, defining expansions like 0.9999... and 1.0000... to be equivalent, and define the arithmetical operations formally. This is equivalent to the constructions by Cauchy sequences or Dedekind cuts and incorporates an explicit modulus of convergence. Similarly, another radix can be used. Weierstrass attempted to construct the reals but did not entirely succeed. He pointed out that they need only be thought of as complete aggregates (sets) of units and unit fractions.

This construction has the advantage that it is close to the way we are used to thinking of real numbers and suggests series expansions for functions. A standard approach to show that all models of a complete ordered field are isomorphic is to show that any model is isomorphic to this one because we can systematically build a decimal expansion for each element.

Construction Using Hyperreal Numbers

As in the hyperreal numbers, one constructs the hyperrationals $^*Q$ from the rational numbers by means of an ultrafilter. Here a hyperrational is by definition a ratio of two hyperintegers. Consider the ring B of all limited (i.e. finite) elements in *Q. Then B has a unique maximal ideal I, the infinitesimal numbers. The quotient ring B/I gives the field R of real numbers.

Note that B is not an internal set in *Q. Note that this construction uses a non-principal ultrafilter over the set of natural numbers, the existence of which is guaranteed by the axiom of choice.

It turns out that the maximal ideal respects the order on *Q. Hence the resulting field is an ordered field. Completeness can be proved in a similar way to the construction from the Cauchy sequences.

Construction from Surreal Numbers

Every ordered field can be embedded in the surreal numbers. The real numbers form a maximal subfield that is Archimedean (meaning that no real number is infinitely large). This embedding is not unique, though it can be chosen in a canonical way.

Construction from Z (Eudoxus Reals)

A relatively less known construction allows to define real numbers using only the additive group of integers $\mathbb{Z}$ with different versions. The construction has been formally verified by the IsarMathLib project. Shenitzer and Arthan refer to this construction as the *Eudoxus reals*.

Let an almost homomorphism be a map $f:\mathbb{Z}\to\mathbb{Z}$ such that the set $\{f(n+m)-f(m)-f(n):n,m\in\mathbb{Z}\}$ is finite. We say that two almost homomorphisms f,g are almost equal if the set $\{f(n)-g(n):n\in\mathbb{Z}\}$ is finite. This defines an equivalence relation on the set of almost homomorphisms. Real numbers are defined as the equivalence classes of this relation.

To add real numbers defined this way we add the almost homomorphisms that represent them. Multiplication of real numbers corresponds to composition of almost homomorphisms. If $[f]$ denotes the real number represented by an almost homomorphism f we say that $0\leq[f]$ if f is bounded or f takes an infinite number of positive values on $\mathbb{Z}^+$. This defines the linear order relation on the set of real numbers constructed this way.

Properties

Completeness: A main reason for using real numbers is that the reals contain all limits. More precisely, every sequence of real numbers having the property that consecutive terms of the sequence become arbitrarily close to each other necessarily has the property that after some term in the sequence the remaining terms are arbitrarily close to some specific real number. In mathematical terminology, this means that the reals are complete (in the sense of metric spaces or uniform spaces, which is a different sense than the Dedekind completeness of the order in the previous section). This is formally defined in the following way:

A sequence (x_n) of real numbers is called a *Cauchy sequence* if for any $\varepsilon > 0$ there exists an integer N (possibly depending on ε) such that the distance $|x_n - x_m|$ is less than ε for all n and m that are both greater than N. In other words, a sequence is a Cauchy sequence if its elements x_n eventually come and remain arbitrarily close to each other.

A sequence (x_n) *converges to the limit* x if for any $\varepsilon > 0$ there exists an integer N (possibly depending on ε) such that the distance $|x_n - x|$ is less than ε provided that n is greater than N. In other words, a sequence has limit x if its elements eventually come and remain arbitrarily close to x.

Notice that every convergent sequence is a Cauchy sequence. The converse is also true:

Every Cauchy sequence of real numbers is convergent to a real number.

That is, the reals are complete.

Note that the rationals are not complete. For example, the sequence (1, 1.4, 1.41, 1.414, 1.4142, 1.41421, ...), where each term adds a digit of the decimal expansion of the positive square root of 2, is Cauchy but it does not converge to a rational number. (In the real numbers, in contrast, it converges to the positive square root of 2.)

The existence of limits of Cauchy sequences is what makes calculus work and is of great practical use. The standard numerical test to determine if a sequence has a limit is to test if it is a Cauchy sequence, as the limit is typically not known in advance.

For example, the standard series of the exponential function

$$e^x = \sum_{n=0}^{\infty} \frac{x^n}{n!}$$

converges to a real number because for every x the sums

$$\sum_{n=N}^{M} \frac{x^n}{n!}$$

can be made arbitrarily small by choosing N sufficiently large. This proves that the sequence is Cauchy, so we know that the sequence converges even if the limit is not known in advance.

"The Complete Ordered Field"

The real numbers are often described as "the complete ordered field", a phrase that can be interpreted in several ways.

First, an order can be lattice-complete. It is easy to see that no ordered field can be lattice-complete, because it can have no largest element (given any element z, $z + 1$ is larger), so this is not the sense that is meant.

Additionally, an order can be Dedekind-complete, as defined in the section Axioms. The uniqueness result at the end of that section justifies using the word "the" in the phrase "complete ordered field" when this is the sense of "complete" that is meant. This sense of completeness is most closely related to the construction of the reals from Dedekind cuts, since that construction starts from an ordered field (the rationals) and then forms the Dedekind-completion of it in a standard way.

These two notions of completeness ignore the field structure. However, an ordered group (in this case, the additive group of the field) defines a uniform structure, and uniform structures have a notion of completeness (topology); the description in the section Completeness above is a special case. (We refer to the notion of completeness in uniform spaces rather than the related and better known notion for metric spaces, since the definition of metric space relies on already having a characterisation of the real numbers.) It is not true that R is the *only* uniformly complete ordered field, but it is the only uniformly complete *Archimedean field*, and indeed one often hears the phrase "complete Archimedean field" instead of "complete ordered field". Every uniformly complete Archimedean field must also be

Dedekind-complete (and vice versa, of course), justifying using "the" in the phrase "the complete Archimedean field". This sense of completeness is most closely related to the construction of the reals from Cauchy sequences (the construction carried out in full in this article), since it starts with an Archimedean field (the rationals) and forms the uniform completion of it in a standard way.

But the original use of the phrase "complete Archimedean field" was by David Hilbert, who meant still something else by it. He meant that the real numbers form the *largest* Archimedean field in the sense that every other Archimedean field is a subfield of R. Thus R is "complete" in the sense that nothing further can be added to it without making it no longer an Archimedean field. This sense of completeness is most closely related to the construction of the reals from surreal numbers, since that construction starts with a proper class that contains every ordered field (the surreals) and then selects from it the largest Archimedean subfield.

Advanced Properties

The reals are uncountable, that is, there are strictly more real numbers than natural numbers, even though both sets are infinite. In fact, the cardinality of the reals equals that of the set of subsets (i.e., the power set) of the natural numbers, and Cantor's diagonal argument states that the latter set's cardinality is strictly bigger than the cardinality of N. Since only a countable set of real numbers can be algebraic, almost all real numbers are transcendental. The non-existence of a subset of the reals with cardinality strictly between that of the integers and the reals is known as the continuum hypothesis. The continuum hypothesis can neither be proved nor be disproved; it is independent from the axioms of set theory.

As a topological space, the real numbers are separable. This is because the set of rationals, which is countable, is dense in the real numbers. The irrational numbers are also dense in the real numbers, however they are uncountable and have the same cardinality as the reals.

The real numbers form a metric space: the distance between x and y is defined as the absolute value $|x-y|$. By virtue of being a totally ordered set, they also carry an order topology;

the topology arising from the metric and the one arising from the order are identical, but yield different presentations for the topology – in the order topology as intervals, in the metric topology as epsilon-balls. The Dedekind cuts construction uses the order topology presentation, while the Cauchy sequences construction uses the metric topology presentation. The reals are a contractible (hence connected and simply connected), separable and complete metric space of Hausdorff dimension 1. The real numbers are locally compact but not compact. There are various properties that uniquely specify them; for instance, all unbounded, connected, and separable order topologies are necessarily homeomorphic to the reals.

Every nonnegative real number has a square root in R, although no negative number does. This shows that the order on R is determined by its algebraic structure. Also, every polynomial of odd degree admits at least one real root: these two properties make R the premier example of a real closed field. Proving this is the first half of one proof of the fundamental theorem of algebra. The reals carry a canonical measure, the Lebesgue measure, which is the Haar measure on their structure as a topological group normalised such that the unit interval [0,1] has measure 1. There exist sets of real numbers that are not Lebesgue measurable, e.g. Vitali sets.

The supremum axiom of the reals refers to subsets of the reals and is therefore a second-order logical statement. It is not possible to characterize the reals with first-order logic alone: the Löwenheim–Skolem theorem implies that there exists a countable dense subset of the real numbers satisfying exactly the same sentences in first order logic as the real numbers themselves. The set of hyperreal numbers satisfies the same first order sentences as R. Ordered fields that satisfy the same first-order sentences as R are called nonstandard models of R. This is what makes nonstandard analysis work; by proving a first-order statement in some nonstandard model (which may be easier than proving it in R), we know that the same statement must also be true of R. The field R of real numbers is an extension field of the field Q of rational numbers, and R can therefore be seen as a vector space over Q. Zermelo–Fraenkel set theory with the axiom of choice guarantees the existence of a basis of this vector space: there exists a set B of real numbers such that every real number can be written uniquely as a finite linear combination of elements of this set, using rational coefficients only, and such that no element of B is a rational linear combination of the others. However, this existence theorem is purely theoretical, as such a base has never been explicitly described.

The well-ordering theorem implies that the real numbers can be well-ordered if the axiom of choice is assumed: there exists a total order on R with the property that every non-empty subset of R has a least element in this ordering. (The standard ordering $\leq$ of the real numbers is not a well-ordering since e.g. an open interval does not contain a least element in this ordering.) Again, the existence of such a well-ordering is purely theoretical, as it cannot be explicitly described.

"Reals" in Set Theory

In set theory, specifically descriptive set theory, the Baire space is used as a surrogate for the real numbers since the latter have some topological properties (connectedness) that are a technical inconvenience. Elements of Baire space are referred to as "reals".

Real Numbers and Logic

The real numbers are most often formalized using the Zermelo–Fraenkel axiomatization of set theory, but some mathematicians study the real numbers with other logical foundations of mathematics. In particular, the real numbers are also studied in reverse mathematics and in constructive mathematics.

Abraham Robinson's theory of nonstandard or hyperreal numbers extends the set of the real numbers by infinitesimal numbers, which allows building infinitesimal calculus in a way closer to the usual intuition of the notion of limit. Edward Nelson's internal set theory is a non-Zermelo–Fraenkel set theory that considers non-standard real numbers as elements of the set of the reals (and not of an extension of it, as in Robinson's theory).

The continuum hypothesis posits that the cardinality of the set of the real numbers is $\aleph_1$, i.e. the smallest infinite cardinal number after $\aleph_0$, the cardinality of the integers. Paul

Cohen proved in 1963 that it is an axiom independent of the other axioms of set theory; that is, one may choose either the continuum hypothesis or its negation as an axiom of set theory, without contradiction.

Natural Number

In mathematics, the natural numbers are those used for counting ("there are six coins on the table") and ordering ("there are more chairs than tables"). These purposes are related to the linguistic notions of cardinal and ordinal numbers, respectively. A later notion is that of a nominal number, which is used only for naming. There is no universal agreement about whether to include zero in the set of natural numbers: some define the natural numbers to be the positive integers {1, 2, 3,...}, while for others the term designates the non-negative integers {0, 1, 2, 3,...}. The former definition is the traditional one, with the latter definition having first appeared in the 19th century. Some authors use the term *natural number* to exclude 0 and *whole number* to include it; others use *whole number* in a way that includes both 0 and the negative integers, i.e., as an equivalent of the *integer* term.

Properties of the natural numbers related to divisibility, such as the distribution of prime numbers, are studied in number theory. Problems concerning counting and ordering, such as partition enumeration, are studied in combinatorics.

History of Natural Numbers and the Status of Zero

The natural numbers had their origins in the words used to count things, beginning with the number 1.

The most primitive method of representing a natural number is to put down a dot for each object. Later, a set of objects could be tested for equality, excess or shortage, by striking out a dot for each object in the set.

The first major advance in abstraction was the use of numerals to represent numbers. This allowed systems to be developed for recording large numbers. The ancient Egyptians developed a powerful system of numerals with distinct hieroglyphs for 1, 10, and all the powers of 10 up to over 1 million. A stone carving from Karnak, dating from around 1500 BC and now at the Louvre in Paris, depicts 276 as 2 hundreds, 7 tens, and 6 ones; and similarly for the number

4,622. The Babylonians had a place-value system based essentially on the numerals for 1 and 10.

A much later advance was the development of the idea that 0 can be considered as a number, with its own numeral. The use of a 0 digit in place-value notation (within other numbers) dates back as early as 700 BC by the Babylonians, but they omitted such a digit when it would have been the last symbol in the number. The Olmec and Maya civilizations used 0 as a separate number as early as the 1st century BC, but this usage did not spread beyond Mesoamerica. The use of a numeral 0 in modern times originated with the Indian mathematician Brahmagupta in 628. However, 0 had been used as a number in the medieval computus (the calculation of the date of Easter), beginning with Dionysius Exiguus in 525, without being denoted by a numeral (standard Roman numerals do not have a symbol for 0); instead *nulla* or *nullae*, genitive of *nullus*, the Latin word for "none", was employed to denote a 0 value.

The first systematic study of numbers as abstractions (that is, as abstract entities) is usually credited to the Greek philosophers Pythagoras and Archimedes. Note that many Greek mathematicians did not consider 1 to be "a number", so to them 2 was the smallest number.

Independent studies also occurred at around the same time in India, China, and Mesoamerica.

Several set-theoretical definitions of natural numbers were developed in the 19th century. With these definitions it was convenient to include 0 (corresponding to the empty set) as a natural number. Including 0 is now the common convention among set theorists, logicians, and computer scientists. Many other mathematicians also include 0, although some have kept the older tradition and take 1 to be the first natural number. Sometimes the set of natural numbers with 0 included is called the set of whole numbers or *counting numbers*. On the other hand, *integer* being Latin for "whole", the integers usually stand for the negative and positive whole numbers (and 0) altogether.

Notation

Mathematicians use N or $\mathbb{N}$ (an N in blackboard bold, displayed as ! in Unicode) to refer to the set of all natural numbers. This set is countably infinite: it is infinite but countable by definition. This is also expressed by saying that the cardinal number of the set is aleph-null $(\aleph_0)$. To be unambiguous about whether 0 is included or not, sometimes an index (or superscript) "0" is added in the former case, and a superscript "*" or subscript "1" is added in the latter case:

$$\mathbb{N}^0 = \mathbb{N}_0 = \{0,1,2,\ldots\}$$

$$\mathbb{N}^* = \mathbb{N}^+ = \mathbb{N}_1 = \mathbb{N}_{>0} = \{1,2,\ldots\}.$$

Some authors who exclude 0 from the naturals may distinguish the set of nonnegative integers by referring to the latter as the *natural numbers with zero, whole numbers*, or *counting numbers*, denoted W. Others use the notation P for the positive integers if there is no danger of confusing this with the prime numbers. In that case, a popular notation is to use a script P for positive integers (which extends to using script N for negative integers, and script Z for 0).

Set theorists often denote the set of all natural numbers including 0 by a lower-case Greek letter omega: ω. This stems from the identification of an ordinal number with the set of ordinals that are smaller. Moreover, adopting the von Neumann definition of ordinals and defining cardinal numbers as minimal ordinals among those with same cardinality leads to the identity $\mathbb{N}_0 = \aleph_0 = \omega$.

Algebraic Properties

The addition (+) and multiplication (×) operations on natural numbers have several algebraic properties:

- Closure under addition and multiplication: for all natural numbers a and b, both $a + b$ and $a \times b$ are natural numbers.
- Associativity: for all natural numbers a, b, and c, $a + (b + c) = (a + b) + c$ and $a \times (b \times c) = (a \times b) \times c$.
- Commutativity: for all natural numbers a and b, $a + b = b + a$ and $a \times b = b \times a$.
- Existence of identity elements: for every natural number a, $a + 0 = a$ and $a \times 1 = a$.
- Distributivity of multiplication over addition for all natural numbers a, b, and c, $a \times (b + c) = (a \times b) + (a \times c)$

No zero divisors: if a and b are natural numbers such that $a \times b = 0$, then $a = 0$ or $b = 0$.

Bibliography

Abbott, Stephen: *Understanding Analysis*, Springer-Verlag, New York, 2011.

Aliprantis, Charalambos D.; Burkinshaw, Owen: *Principles of Real Analysis,* Academic Press, NY, 2018.

Andrews, G. E. R. Askey, R. Roy: *Special Functions*, Cambridge University Press, London, 2016.

Apostol, T.: *Mathematical Analysis*, Addison-Wesley, London, 2004.

Askey, R. A.; Roy, R.: *NIST Handbook of Mathematical Functions*, Cambridge University Press, London, 2022.

Bartle, Robert G.; Sherbert, Donald R.: *Introduction to Real Analysis,* John Wiley and Sons, New York, 2016.

Bressoud, David: *A Radical Approach to Real Analysis*, MAA, NJ, 2017.

Browder, Andrew: *Mathematical Analysis: An Introduction*, Springer-Verlag, New York, 2016.

Dangello, Frank; Seyfried, Michael: *Introductory Real Analysis*, Brooks Cole, London, 2009.

Dieudonné, Jean: *Foundations of Modern Analysis*, Academic Press, New York, 1990.

Dunford, Nelson; Jacob T., Schwartz: *Linear Pperators. Part I: General Theory*, Wiley-Interscience, New York-London-Sydney, 1998.

Edwin Hewitt and Karl Stromberg: *Real and Abstract Analysis*, Springer-Verlag, New York, 2009.

Emil Artin: *Exposition by Emil Artin: a Selection*; American Mathematical Society, Providence, RI, 2016.

Giaquinta, Mariano; Modica, Giuseppe; Souèek, Jiøí: *Cartesian Currents in the Calculus of Variation I*, Springer Verlag, Berlin-Heidelberg-New York, 2018.

Henrici P.: *Applied and Computational Complex Analysis,* Wiley, NY, 2006.

Hudjaev, Sergei Ivanovich; Vol'pert, Aizik Isaakovich: *Analysis in Classes of Discontinuous Functions and Equations of*

Mathematical Physics, Martinus Nijhoff Publishers, Dordrecht-Boston-Lancaster, 2005.

James D. Bonnar: *The Gamma Function*, CreateSpace Publishing, Seattle, 2023.

Kolmogorov, Andrej N.; Fomin, Sergej V.: *Introductory Real Analysis*, Dover Publications, New York, 1999.

Kreyszig, E.: *Advanced Engineering Mathematics,* Wiley, NY, 2016.

Markushevich, A.I.: *Theory of Functions of a Complex Variable,* Prentice-Hall, NY, 2005.

Milton Abramowitz and Irene A. Stegun: *Handbook of Mathematical Functions with Formulas, Graphs, and Mathematical Tables,* Dover, New York, 2002.

Moreau, Jean Jacques: *Topics in Nonsmooth Mechanics*, Birkhäuser Verlag, Basel-Boston-Stuttgart, 2008.

Nico M. Temme: *"Special Functions: An Introduction to the Classical Functions of Mathematical Physics"*, John Wiley & Sons, New York, 2016.

Press, WH; Teukolsky, SA; Vetterling, WT; Flannery, BP: *Numerical Recipes: The Art of Scientific Computing*, Cambridge University Press, New York, 2017.

Riesz, Frigyes; Szõkefalvi-Nagy, Béla: *Functional Analysis*, Dover Publications, New York, 2010.

Rudin, Walter: *Principles of Mathematical Analysis*, McGraw–Hill, New York, 2018.

Scheidemann, V.: *Introduction to Complex Analysis in Several Variables,* Birkhauser, 2015.

Shaw, W.T., *Complex Analysis with Mathematica,* Cambridge University Press, Cambridge, 2016.

Spiegel, Murray R.: *Theory and Problems of Complex Variables - with an introduction to Conformal Mapping and its Applications,* McGraw-Hill, NY, 1994.

Titchmarsh, E.:*Introduction to the Theory of Fourier Integrals,* Chelsea Pub. Co., New York, 1998.

Walter Rudin: *Real & Complex Analysis*, McGraw-Hill, New York, 2017.

Whittaker, E. T. and G. N. Watson: *A Course of Modern Analysis,* Cambridge University Press, London, 2016.

Index

❑❑❑